HOW NUMBERS LI

HOW NUMBERS LIE

A Consumer's Guide to the Fine Art of Numerical Deception

By Richard P. Runyon

Illustrations by Frank Corbie

THE LEWIS PUBLISHING COMPANY • Lexington, Massachusetts

First Edition

Produced in the United States of America.
Designed by IRVING PERKINS ASSOCIATES.
Published by THE LEWIS PUBLISHING COMPANY, INC., Lexington,
Massachusetts 02173.
Distributed by THE STEPHEN GREENE PRESS, Fessenden Road at Indian Flat,
Brattleboro, Vermont 05301.

Library of Congress Cataloging in Publication Data

Runyon, Richard P.
How numbers lie.

Includes bibliographical references and index.
1. Statistics 2. Statistics—Anecdotes, facetiae,
satire, etc. I. Title.
HA29.R852 001.4'225 80–29415
ISBN 0–86616–000–0
ISBN 0–86616–001–9 (pbk.)

Contents

Preface

The great nineteenth century British Prime Minister, Benjamin Disraeli, once said, "There are three kinds of lies: lies, damned lies, and statistics." Had he been Henry VIII, he might have added, "And the best of them is the worst."

For statistics is a strange bedfellow. It openly consorts with liars, hoodlums, thugs, politicians and other cheaters. Yet behaviorial, social, biological and physical scientists do it great homage. Hardly a learned paper in any of these fields is published without being accompanied by a wealth of statistical information with conclusions buttressed by statistical proof. But their legitimate use is not restricted to the groves of academe. Business executives, financial advisors, stock brokers—indeed the entire spectrum of the work-a-day world—base billion dollar decisions on speculations and facts in which statistics is an essential ingredient of the decision-making soup. Have you ever opened the financial page of your daily newspaper and failed to find any reference to the consumer price index, the cost of living index, or the gross national product? No way.

Why does statistics seduce the crook and the clergy, the shiftless and the diligent, the deceitful and the seeker after truth? Perhaps this may be answered by the well-known aphorism, "Figures don't lie but liars figure." It reminds us that when statistics are used to beguile and to deceive, the fault lies within us rather than in statistics. Is this after all, much different from humanity's use of any great discovery—TNT, atomic energy, cloning, and genetic engineering?

The main purpose of this book is to take some of the mystery out of the use of statistics. It's too important an analytic tool to be left exclusively in the hands of either knaves or professionals. If for no other reason than self protection, the consumer should be aware of both facets of statistical analysis—the legitimate as well as the deceptive applications of this wonderful tool. If, by the end of reading this book, you are able to answer a few of the following questions, I shall be well satisfied I have accomplished my objective.

Why do gambling situations favor the house, the bookie, or the track?

Since all of life's decisions involve risk, are there any standards for evaluating the risk against potential gains?

Why is computer fraud one of the best risk-taking ventures available today?

What sort of chicanery should I look for when examining graphs and charts?

Does any statistic have meaning in the absence of a standard of comparison?

Why are medical statistics so confusing and often contradictory?

Why do so many compounds and treatment procedures "found safe" in the laboratory and on clinical trials suddenly turn into nightmares? Can we really prove anything safe?

One final word—an expression of gratitude to Frank Corbie who did the artwork for this book. Gifted with a tremendous sense of humor and an unerring eye for the incongruous, he has brought many smiles to my face as I wrote this book.

OW NUMBERS LI

CHAPTER 1

From Birth to the Gaming Tables

The year was 1502. A massive, high-pressure front, blasting in from the North Sea, was besieging Antwerp with bone-chilling cold. Domingo shivered in his pad of straw and searched vainly for a twig in hope of breathing a bit of life into the earthen fireplace. For him, the winter's icy breath added but a minor perturbation to his sea of troubles. In his early twenties, unemployed with no job in sight, and down to his last few livres, he had spent a restless night trying to fathom some order in his chaotic life. Tallying his assets, he judged himself bright, imaginative, energetic, articulate, and ambitious. But the debit side seemed to overwhelm the positive—he had little formal education and no marketable skills. The prospects of the life of a common laborer or dockhand held little appeal to him. He turned to his brother, who was just beginning to emerge from sleep.

"It's ironic, Bernardo. Here we are in the greatest center of commerce the world has ever known, and we don't even have a fagot to keep our bodies warm."

"I think we should look at the docks again today," his brother replied sleepily. "Something's bound to open up for us."

"We've been saying that for weeks, and where has it gotten us? Besides, the docks are not for us. There are riches out there, wealth that numbs the mind as surely as the cold numbs the body. There are livres by the thousand waiting to be plucked by bold and venturesome ideas."

"Ideas make money? Come now, has the cold got to your brain? People such as we make money only through the pain of our backs and the sweat of our brows."

"Nonsense, Bernardo. Only fools submit weakly to such a

fate. I have a plan to make us a fortune. I lay awake all night long thinking it through. I tell you it can't miss."

"All right, let's hear it, brother Domingo, but I warn you in advance—I am not likely to make an enthusiastic listener, what with my stomach shrivelled up like an amaryllis bulb prior to planting."

"We'll soon put so much food in your stomach it will look like a ripe melon," bubbled Domingo. "Listen to me. If you were to father a child, what would you want its sex to be?"

"Why, a boy, of course. Everyone wants a boy."

"What would you say if I could guarantee that the child will be whatever sex you chose?"

"Hush with such talk, Domingo. Only God has that power. We'll be burned at the stake as heretics if you don't keep a tighter tongue in your mouth."

"No, you get me wrong, Bernardo. I said we would guarantee the sex of your choice. I never said that we could make good on the guarantee."

"Worse still. You're talking of lying and deceit. Our parents are spinning in their graves at such talk."

"I said we would guarantee the sex of your choice. I never said we could make good on the guarantee."

"Again you misjudge me, brother. First, hear me out and then judge me as a thief or genius, whichever is your wont."

"I doubt the word is genius from what I have heard so far."

"Perhaps yes, perhaps no, just hear me out. Have you ever heard a proud prospective parent speak of a coming child who did not also speak wistfully of the desired sex?"

"Of course, and most want boys because boys become men and men are more valuable than women."

"Not necessarily, Bernardo. Imagine there are two groups stranded on separate but identical island paradises. Imagine next that one consists of a single man and nine women and the other of nine men and one woman. Which would plenish their island more rapidly? The one with the single male, of course. This proves that, in matters of propagation, the male is more dispensable than the female. One fertile male can serve as well as nine, but one woman could never substitute for as many females. It is for this reason, I think, that males labor in all the dangerous professions. You will never see a woman go to war, not because of a physical weakness, as is often stated, but because she is too valuable to risk losing. But males are a livre a dozen.

"But I stray. My scheme is simple enough. We charge a fee to guarantee the sex of a child. If the parents wish a boy, we shall guarantee the birth of a child of like sex for a fee of, say, forty-eight livres. When a boy is born, everybody is happy. The parents have a child for whom a name has already been chosen and we have forty-eight livres free and clear."

"Anyone would be an idiot to agree to such a proposition. If a girl is born, they lose all."

"Ah, that is where my plan is ingenious. In the event we fail to deliver the sex the parents wished, we return their forty-eight livres and throw in an additional thirty. Thus, the parents have some solace for their loss. They also wind up with a total of seventy-eight livres, more than enough to pay for the midwife and a week's supply of Pampers."

"Pampers?"

"Ah, yes. That is a marvelous disposable diaper that they will invent in the twentieth century. But again I stray. Think of the arithmetic of this plan. Let us say we negotiate one hundred contracts. We collect a total of four thousand eight hundred livres. If our magic powers fail us completely, as I suspect they will, we will be correct about half the time, and on the average wrong half the time. We shall have to pay out seventy-eight livres to fifty pairs of parents, give or take a few. Seventy-eight times fifty is three thousand nine hundred livres; four thousand eight hundred minus three thousand nine hundred leaves us nine hundred livres. Not bad for just shuffling papers."

"Not bad, but what will you say before the confessional?"

"I'm afraid I lose your drift."

"It's dishonest. That's what I am saying, Domingo. It is a scam, pure and simple."

"Scam?"

"You're not the only one in touch with the future. 'Scam' is a twentieth-century word for a ripoff."

"A ripoff?"

"Yes. Something the oil companies will someday do."

"Oh. But this is totally different, my dear brother. How can anything be wrong, dishonest, or bad if it satisfies everyone?"

"Yes, but I think it is gambling. And gambling, except for bingo, is a sin."

"Yes, it does appear to be gambling. Ergo, we'll have to give it a respectable name so it won't be gambling any more. We'll call it *assurance . . . birth assurance*."

"Birth assurance?"

"Yes. For a fee we shall assure the assured of a particular sex of a newborn child. If we deliver, we retain the fee. If we fail, we enrich the lives of the assured. Bernardo, I think I am on the edge of something big. Why, someday I see the possibility of life assurance companies."

"Oh, come now. While I see the merit of birth assurance, people would never be so naive as to take out assurance on their own lives. They can't take it to the grave, you know."

"Of course not. But they go to the grave knowing they have not left their loved ones in poverty. You see, upon the death of the assured, the assurance company would pay those who benefit—oh, we'll call them beneficiaries—some agreed-upon sum of money."

"Where will the assurance company get these funds?"

"Why, they'll have been using your money for investments. They'll invest in ships, real estate, munitions, and all those other wonderful human endeavors we call commerce. They'll use your money for their profit. How beautiful a scheme. And when they pay off, years later in most cases, it will be in inflated livres."

"Oh, yes. I forgot about those inflated livres. You give them use of the livres when it is worth a livre and they pay your beneficiary back when a livre is not worth a livre any more."

The Rich Get Richer

The above was a somewhat fictionalized account of a conversation between Domingo Symon Maiar and his brother Bernardo. A contract between them and two women survived until the present century.[1] In this contract, they agreed to pay each woman thirty livres if the respective offspring was a girl, but they were to receive forty-eight livres for the birth of a son.[2] We don't know what happened to Domingo and Bernardo. However, we do know that their chances of setting up a successful brokerage firm depends on their financial status to begin with. If poor, their chances are drastically reduced. Why?

Let's assume that the two brothers began their business with exactly sixty livres. What are the prospects that their two contracts will leave them in the black? Let's assume that the birth of a girl and/or boy are equally likely. It's like tossing a coin—heads, it's a girl; tails, it's a boy. We may then calculate the various possible outcomes and their net return. These are shown in table 1.1.

Table 1.1

		First Woman	
		Boy (+48)	**Girl (−30)**
	Girl (−30)	48 − 30 = 18 (gain) A	0 − 60 = −60 (loss) B
Second Woman	**Boy (+48)**	96 − 0 = 96 (gain) C	48 − 30 = 18 (gain) D

[1] Hogben, L., *Mathematics for the Millions* (N.Y.: W. W. Norton, 1940) p. 577.
[2] Ibid.

Note that if the first woman has a boy and the second a girl, (cell A), our young entrepreneurs have increased their holdings by eighteen livres. The same is true if the first woman produces a girl and the second a boy (cell D). Nirvana is reached in cell C, where the birth of two boys leads to a gain of ninety-six livres. But then there is cell B—total loss, a wipeout, bankruptcy. In the situation described here, the chances are about one in four that this will occur.

Well, that's not bad, you say. Three outcomes provide gains and only one a loss. You wish you had such good odds when you place a bet on the New Jersey football Giants! But these two selections are quite different. You hope, when you place that small bet on your favorite football team (or buy a lottery ticket; spend a weekend in Atlantic City or Vegas; go on a bingo binge; etc.), the outcome is not a matter of economic survival. If it is, you are in big trouble. You simply are not in a position to sustain a run of bad luck. And such runs will occur predictably and with regularity. Let us suppose that by carefully husbanding their winnings Domingo and Bernardo managed to amass a small fortune—300 livres. Confidentially, they accepted ten more contracts. Then disaster struck. Ten consecutive losses! Admittedly, such outrageous fortunes are quite rare—slightly less than one in a thousand, but they do occur with disgusting regularity. The difference between the rich and the less affluent is that the moneybags can sustain long runs of bad luck, whereas we common folk are brought to our knees by prolonged adversity. I have never heard of a major oil company bellying up as a result of twenty consecutive dry holes. Do you know of any wildcatter who can survive more than two or three?

Because of this built-in protection against adversity, the affluent can afford to put themselves into a greater number of territories that involve financial risk. Moreover, the crushing reality is that the risk can be so small as to be almost nonexistent. Suppose, for example, that a wealthy American were to revive the sixteenth-century scam of the Maiar brothers. For the purposes of simplification, we previously assumed that the likelihood of a male and female birth are equal. This is not the case. The proportion of males born to white parents in the United States has averaged about .5141 over recent years. For blacks, this proportion is somewhat lower—.5069. Moreover, these pro-

Table 1.2 Proportion of male births, by race, in the U.S. between 1971–1977. Note that the proportion of boys is consistently above .50 year after year, and quite stable for both whites and blacks. The proportion of boys born to blacks, however, is consistently lower than those born to whites.

Year	White	Black
1970	.5143	.5076
1971	.5136	.5069
1972	.5139	.5059
1973	.5138	.5068
1974	.5143	.5074
1975	.5143	.5074
1976	.5141	.5067
1977	.5141	.5064
Average	.5141	.5069

portions of male births have remained remarkably stable from year to year, as can be seen in table 1.2.

Here's what a well-heeled financial operation can afford to do. It can guarantee a male birth, *even odds*. If the bet is $100, the parents pay $100 if a boy is born and collect $100 for the birth of a female. If 100,000 bucks are placed each day, the average daily profit would be $2,820, if betting is restricted to white males.[3] It would be somewhat lower if bets were taken irrespective of race. Note that the financial backers would have a negative cash flow on any given day only if the number of female births exceeded the number of male births on that day. Knowing that there are approximately 7,400 births per day within our country at the present time, we can ascertain that the likelihood of female births outnumbering male births on any given day is less than one in a hundred. What is more to the point, if we assume that 740 sets of parents place bets daily, a negative cash flow on any given day will occur only about 22 percent of the time. And 78 percent of the time the cash flow will be positive. Of course, there will be days when the financial backers will take a bath. But, unlike us common folk, the rich people can take temporary setbacks in stride, just as a gambling casino is not plunged into panic and despair every time someone

[3] The wins from male births would average $100,000 × .5141 = $51,410. The losses from female births would be $100,000 × .4859 = $48,590. The difference between the two represents the average gross daily profit.

breaks the bank. In fact, occasional big winners are good business for any gambling enterprise. They get a lot of free media coverage.

Of course, there will also be days of larger-than-average profits. But these daily fluctuations do not excite the gambling connoisseur (be he or she common gambler, stock market broker, dealer in the commodity exchange, spot buyer of oil in the Amsterdam market, buyer for a department store, burglar, etc., *ad infinitum ad nauseam*). What is important is the bottom line—the end-of-the-year financial statement. In the little hypothetical scam we have been entertaining here, the profit would be slightly in excess of 1,000,000 smackeroos, minus graft and other overhead costs. Not bad for a virtually risk-free operation, *n'est-ce pas?*

Thus do we see in this homey example that it takes money to make money and there's good reason why the rich get richer.

Let's face it. Life is a gamble. Virtually every situation involves some element of risk as well as some potential gain. Only one thing appears certain. There is no certainty. Not even the most sophisticated statistical analysis can remove all elements of risk. At best, they can define and circumscribe the risk and provide valuable guidelines to sources of action that maximize potential gain.

From Birth to the Gaming Tables

The baby-betting scam, although hypothetical, illustrates several important principles of risk taking:

Rule 1: A venture is worth consideration only if the potential gain exceeds the potential long-term losses.

Rule 2: There must be sufficient resources to withstand the statistically inevitable short-term reversals.

Rule 3: The gain per transaction need not be great as long as there is a sufficiently rapid turnover rate.

What better place to demonstrate these principles than on the gaming tables. (See box 1.1.) Contrary to popular opinion, gaming houses are scrupulously honest. They must be. It would be sheer stupidity to take the risk of being caught when, as we shall see, profits are assured.

Box 1.1

"Excuse the interruption. Let me introduce myself. I'm Hal Fast, a friend, advisor, and mentor to the author of this book. To tell you the truth—and don't whisper this to a soul—he's the writer and I'm the brains of this outfit."

"Now, wait a minute. I heard every word you said."

"Eavesdropper!"

"You said it loud enough to be heard in the next county."

"So what? Every word is the truth. Like Howard Cosell, I tell it like it is."

"I don't care. I hired you for one thing only."

"You don't care about the truth?"

"Another word from you and it's Three strikes, you're out. Do I make myself clear?"

"Perfectly clear."

"So why did you interrupt? What did you have to say?"

"Nothing."

"Come on, now, let's not pout. I hired you to do the color commentary for this book and I expect you to act like a professional."

"Nobody prevents Howard from saying how good *he* is."

"OK. Let me say it for you. You're great. The greatest. That's why I hired you."

"Thank you. But that's not the way you put it. You said you

"I'm Hal Fast, a friend, advisor, and mentor to the author of this book. He's the writer, and I'm the brains of this outfit."

wanted me to spice up the dull parts of the book. Stop frowning. I won't tell a soul."

"I'm counting on it. Now, what were you going to say?"

"Some of your readers might think it strange that a gambling casino is so concerned about the honesty of its equipment."

"I imagine so. I was even a bit surprised that I wrote it."

"It's not because they're churchgoing pillars of the community, you know."

"I suspected as much."

"It's just that they have so much to lose if the equipment goes sour."

"How come?"

"Their profit is built into their equipment and games by the laws of chance. They know precisely what their profit per transaction will average. It's called vigorish, you know."

"What's called vigorish?"

"Their profit per transaction. It's from the Russian *vyigrysh*, meaning *winnings* or *profit*."

"Hal Fast, you're wandering."

"Oh, yes. As I said. They know, almost to the penny, how much they'll win out of every dollar bet. They don't welcome the introduction of uncertainty."

"Such as a flat bearing?"

"Or a die that is slightly worn on one side. Anything that changes the odds is like the plague. To paraphrase the bard, they'd rather bear the riches they have than fly to others they know not of. Besides, it is a veritable certainty that someone in the casino is keeping a record of how every die, every roulette wheel, every one-armed bandit is performing. As soon as a flaw is detected, he'll start betting the flaw. He could make a bundle before anyone is the wiser. That's why the dice are changed many times a day, and the equipment is kept well oiled. Casino owners are very conservative people. They don't like surprises."

"And they are scrupulously honest?"

"Yes, the operation of a gambling casino is one activity about which I can say without reservation, 'Honesty is the best and the winning policy.' "

Take the roulette wheel. It has thirty-eight equally spaced openings numbered 00, 0, 1, 2, . . . , 35, 36. The gambler, let's call her Borne L. Oser, places a bet, say $1, on any number. The croupier starts the wheel in motion and drops a ball into it while it is spinning. If the ball winds up in the opening on which Borne bet, she wets in her pants a little and collects $35. Not bad for one spin of the wheel. But here's the rub: If the ball lands in any other opening, she pays the house a buck. On the surface, the prospects don't sound all that bad. The risk is $1 and the payoff is $35. Let's look at what we may expect her to do over the long run. Let's say she places 100 consecutive bets. Is she likely to come out a winner?

On the average, we would expect her to win once out of every 38 spins of the wheel. In every 100 spins, then, she would win an average of $1/38 \times 100 = 2.6316$ times. Each win is worth $35, so her total winnings per 100 spins would be 35×2.6316 = $92.11. In the meantime, she would lose, on the average, 97.3684 times in 100. At a buck a loss, she would part with $97.38 per 100 spins. The difference ($92.11 − $97.37) represents a loss of $5.26 per 100 spins, or about a nickel a spin. Given "average" luck, her nest egg of $100 should carry her through about 2,000 spins. But note: Borne L. Oser is a loser in the long run. She might occasionally have a good night and come out a winner, but she is sure someday to sing the gambler's lament "Brother can you spare a hundred bucks?[4]—I feel a hot streak coming on!"

For the casino's part, whether or not *she* wins is immaterial. The management knows that every dollar bet will return an average of a little more than a nickel. If you were an oil company executive, you might bemoan such a low rate of return (see box 1.2). But the proprietor of a casino knows better. It's not the rate of return that's important, but the rate of dollar turnover. If a roulette wheel can turn over $100,000 in a twenty-four-hour period, it will return the casino an average gross daily profit of $5,260. That's the bottom line, baby, not who wins or loses. So don't go to Vegas, Atlantic City, Puerto Rico, or Monaco with the idea of making a killing. Just be happy if you escape with a Sheraton-Hilton bath towel.

A demonstration by the famous Harvard psychologist B. F. Skinner beautifully illustrates the value of a high rate of turnover. He purposely rigged some one-arm bandits to deliver a greater proportion of wins as money was put in faster. Then he

[4] Substituted for a dime as an inflationary adjustment.

Box 1.2

Those Wonderful Oil Companies and Their Flying Statistics

"It's funny you should say that," says Professor Fast. "I just read this ad in the morning paper. It's two thirds of a page and is headed 'Chevron Energy Report.' Take a look at it. Just as you say, at least one oil company seems rather unhappy about the fact that it is only average on profits per dollar of sales. But is this a valid basis for comparison?

> The *Arizona Daily Star,* March 25, 1980. Section C, p. 3.
>
> Chevron energy report: Compared to all U.S. industry—Chevron's nickel profit makes us just average.
>
> The average profit for all major U.S. industries last year was 5.5¢ on a sales dollar.
>
> By comparison, in 1979 Chevron made about 5.1¢ on each sales dollar of U.S. petroleum sales—a little less than the average of U.S. industry.
>
> Even on our worldwide sales, we still made less on a sales dollar than the average of all U.S. industries.
>
> Like most companies, we reinvest most of our worldwide profits after dividends plus cash from operations (including depreciation). In 1980, Chevron's reinvestment in energy development in the U.S. will be a *record* for us—more than *twice* our '79 U.S. profit.
>
> Investment in U.S. energy development is the best way to help move America toward energy independence. But, we must all continue to conserve as much energy as possible.

"Is nothing to be said about the rate of turnover of each dollar of sales? Think about it awhile. In this country, we presently use over seven hundred million gallons of black gold a day. If each gallon sells at a price of about one dollar, over seven hundred million dollars of petroleum sales, on the average, turn over each day. If all oil companies are making about five cents per dollar of sales, the daily profit is a niggardly thirty-five million dollars. (Why that's more than I make in six months!) In a year, it's only about thirteen billion dollars PROFIT!

"Except for gambling operations, I can think of no business with such a magnificent rate of sales turnover.

"How long does a non-oil merchant keep an item in inventory before he turns the dollar over? A month, six months, a year? Isn't the sale of oil in many ways like the operation of a gigantic discount house? Less profit is made per dollar of sales but, oh! that turnover rate, that beautiful sales volume!"

provided some funds to undergraduate students to blow in the infernal machines. The students on the rigged machines put money in like crazy, won a lot more often, but were the first to go broke. It's an elemental fact. A 2 percent rate of loss can amount to much more than a 4 percent loss within the same time period if the 2 percent is being lost three times as fast. It's also elemental that if you are the house you'll do everything possible to encourage fast betting. "Have another drink, buddy. Maybe it will put Lady Luck on your side." Don't bet on it!

Don't Bet on the Nags Either

Theoretically, it should be possible to make a buck on the ponies. If the race is honest and you are blessed with the wisdom of a handicapper, you should have a definite advantage over the superstitious run-of-the-mill, chain-smoking, hunch-taking gamblers. The situation is totally unlike the gaming casinos, where the odds are built into the games of chance with mathematical precision. In horse racing, the *odds are established by the bettors themselves*. If a given nag is a favorite with the crowd, many people place their bets on its nose. The more money bet, the lower the odds. In other words, if you win, the payoff is not very great. Betting in such a way violates the first rule of a favorable risk-taking situation: A venture is worthy of consideration only if the potential gain exceeds the potential long-term losses. If a horse is likely to win in four out of every five races it runs and you take even odds, you are destined to be burned in the long run if you persist in taking this foolish risk. In fact, your average loss per two dollar bet would be forty cents.

There is a fact about the psychology of betting that, theoretically at least, opens the door to a positive cash flow. Research has established that the gambling public tends to overbet both the favorite and the long shots. As a result, the odds are lowered and the payoff is less than it should be. Let's illustrate with the following scenario of despair.

Imagine you are at the racetrack looking gloomily at the last two bucks to your name. What should you do? Hold on to it so you have bus fare to the pawnbroker? Bet it on the favorite? On an also-ran?

Box 1.3 Beating the Point Spread

Did you ever consider betting on your favorite football team? Mystified by the point spread? I prevailed on Hal Fast to give us the inside scoop. He explained as follows.

Here's the deal. Chief Warhorse has just collected his weekly wampum of $500. Since the money is burning a hole in his wampum belt, he frantically searches out his bookie, Honest John, before the loot begins to sear his flesh.

"What's the point spread on the Kansas City / Pittsburgh game?" he asks.

"The Chiefs and seven points," John replies.

"Good. Put five hundred dollars on the Chiefs."

This means that Chief Warhorse is in the chips if one of two things happens: The Chiefs win, or the Steelers win by less than seven points. If the Steelers win by exactly seven points, it is a standoff. Neither the chief nor the bookie collects.

Sally Kaz is hot for the Steelers. She takes her five hundred dollars to Honest John and gets the following deal: Take the Steelers and give up twelve points. If the Steelers win by thirteen or more points (which Sally assumes is a lead-pipe cinch), she's in clover. If they win by exactly twelve points, it's a stand-off. If the Chiefs win ("No way!" says Sally) or if Pitt wins by four field goals or less, Sally drops the bundle.

This is what Honest John's book looks like in the transaction.

Outcome of the Game	Net Outcome of Transaction
1. Steelers win by 13 or more points. Pay Sally $500. Collect $500 from chief.	00
2. Steelers win by 12 points. Stand-off with Sally. Reduce chief's wampum by 500 bucks.	+500
3. Steelers win by 8 to 11 points. Collect $500 from both suckers.	+1,000
4. Steelers win by exactly 7 points. Stand-off with chief. Collect $500 from Sally.	+500

5. Steelers win by less than 7 points. Pay chief $500 and collect same from Sally.	00
6. Chiefs win. Double chief's wampum holdings; double his delight! Sally loses both ways. Gives up $500 and quaffs the bitter dregs of defeat.	00

Did you notice something about the entries in the column labelled "Net outcome of the transaction?" Absolutely, you're right. *Honest John cannot lose.* As long as he lays out his bets in an intelligent manner, the worst disaster that can befall him is to break even. Notice something else. He stands to make his biggest killing when the final result is between the two point spreads. He collects from everybody whenever the Steelers win by more than seven and less than twelve points.

Have you ever noticed that the major betting scandals have usually involved point shaving? Why is this so? Simple enough. The gambling interests want the final score to be in that hallowed ground between point spreads. So you tell the star player most in control of the final outcome, "It's not as if we're asking you to lose. We're offering you fifty g's just to shave a few points off the winning margin. That way, nobody's the wiser and we're all winners."

All winners, that is, but the poor sucker—the bettor.

You look at the morning line and see that Trois Jambes, entered in the third race, carries odds of 99–1. You have three names (first, middle, and last), you have had three spouses, and it is the third day of the third month of the year. It can't miss! All the favorable signs are there. Incidentally, it is important to note that the morning line fairly accurately reflects the probabilities of winning. Thus, Trois Jambes might be expected to win about one time in a hundred, if its owner can resist the temptation to donate the nag to a dog-food entrepreneur and claim the donation as a charitable deduction. You quickly calculate the payoff. Risking only $2, you stand to win $198. That, plus the original $2 grubstake, will leave you with a grand total of $200. Why that's almost enough to fill the gas tank with unleaded! With a new surge of enthusiasm, you place the bet and then watch the tote board. A creepy-crawly feeling settles in your abdomen. The odds, starting out at 99–1 begin to inch down. Apparently other astute observers like yourself have noted the confluence of threes. Bad cess to them.

By post time, the odds have lowered to 20–1. If your nag wins, you'll wind up with $42 rather than $200. Well, enough for a sumptuous meal of a hamburger and Coke, extra large with 60 percent ice. You look at your horse for the first time.

"Why, it has only three legs," you gasp.

"*Naturellement*," a bystander agrees, "That's *précisement* what its name means—three legs."

Naturellement, you lose.

But deep down, didn't you know that this would happen? Hasn't it happened to you before?

Now let's look at another scenario—one that offers the possibility of a positive cash flow. You stand quietly and unobtrusively near the betting window, one eye on the morning line and the other on the tote board. You note with satisfaction that, as expected, the odds on both the favorites and the long shots are dropping. But something else is also happening. Very significant. *The odds are going up on the neutral horses*—neither favorite nor long shot, the morning line odds on Fair-to-Middlin were 6–1. Its chances of winning, then, are about one in seven. But, because relatively few people bet on this nag, the tote boards read eight or nine to one. If we can grant a whole bunch of assumptions (the race is honest, the morning line is a fairly accurate reflection of the probability that the horse will win, etc., etc., etc.), we should be able to do pretty well in the long run. For example, if we bet two bucks on seventy-seven races in which the morning-line odds were 6–1 and the actual track payoff was 9–1, we would expect, on the average—

1. to win eleven races for a payoff of $198 ($2 × 11 × 9);
2. to lose sixty-six races for a payout of $132 ($2 × 66).

Our expected gain would be $66, give or take a few. Once again, a vital caveat (rule 2): There must be sufficient resources to withstand the statistically inevitable short-term reversals. About 46 percent of the time you will sustain five consecutive losses and ten consecutive losses about 21 percent of the time. Why, even twenty losses in a row will occur almost 5 percent of the time! (See table 7.1, row 6:1.) Need I remind you that the cards are stacked against the beggarly risk-taker?

And there's a rub. There's always a rub. The track computer calculates the odds only after the track and the state have taken their profit. Thus, the odds on the tote board are always lower than they would have been if these two culprits had kept their grasping hands out of the entire transaction. Because of them, it might be tough to find a change from the morning line to post time sufficient to justify the financial risk.

"What about going to a private source?" you ask.

"To what?"

"You know, a bookie. At least, the track and state don't have their hands in the till."

"That's true. But he gives track odds. That means the computer is working for him, too."

So Who's the Sucker? I'll Let You Decide.

Let's review the three rules of a favorable-risk venture and see how the track (and state) make out vis-à-vis the gambler.

Rule 1: The venture is worthy of consideration only if the potential gain exceeds the long-term losses.

The track cannot lose on the gambling operation, as such, short of a computer failure. A percentage of the handle is deducted before the race is run. However, it does have an enormous overhead. It would lose if people stopped going to the track and betting on the ponies. Wanna bet this doesn't happen?

The gambler who plays hunches is preordained to suffer grief. A system based on the psychology of the bettor shows some possibility of success if a sufficiently wide disparity can be found between the odds at post time and the morning line odds.

Rule 2: There must be sufficient resources to withstand the statistically inevitable short-term reversals. In its gambling operation, the track cannot suffer short-term reversals, since the profit is deducted beforehand.

The gambler, even if functioning with a workable system, will experience numerous short-term "dry holes." The greater the

Table 1.3 The probability of obtaining selected numbers of *consecutive losses* at varying odds from 2:1 to 10:1. Note that the higher the odds the more likely you are to experience a run of "bad luck." For example, at 10:1, the chances are almost 50:50 that you will suffer 8 consecutive losses. The probability is even quite high (.15) that you will lose 20 in a row.

	Number of Consecutive Losses							
Odds	**2**	**5**	**8**	**10**	**12**	**15**	**17**	**20**
2:1	.44	.13	.04	.02	.008	.002	.001	.0003
3:1	.56	.24	.10	.06	.03	.01	.008	.003
4:1	.64	.33	.17	.11	.07	.04	.02	.01
5:1	.69	.40	.23	.16	.11	.06	.05	.03
6:1	.73	.46	.29	.21	.16	.10	.07	.05
7:1	.77	.51	.34	.26	.20	.13	.10	.07
8:1	.79	.55	.39	.31	.24	.17	.14	.09
9:1	.81	.59	.43	.35	.28	.21	.17	.12
10:1	.83	.62	.47	.39	.32	.24	.20	.15

post-time odds, the more likely he or she will encounter a run of "bad luck." Table 1.3 shows the probability of experiencing selected runs of bad luck at varying odds from 2–1 to 10–1.

Rule 3: The gain per transaction need not be great as long as there is a sufficiently rapid turnover rate.

The track and the state need worry only about the total handle. The rate is fixed by law. A large handle means big profits, a small handle, smaller profits. If one million bucks are bet on a given day and the percentage allocated to the track and state is 15 percent, the gross profit would be $150,000 (not including gate receipts and concessions).

If a gambler has devised a system that provides a long-term positive cash flow no matter how small, he or she should celebrate the fact and not be concerned or lose sleep over the rate of return per transaction. Very few gamblers make a living off the track.

Box 1.4 Going One on One

"You can say that again." It's Hal Fast speaking. He explains the situation as follows.

If you must gamble, your best chance is to go one-on-one. Let me explain. Wherever there are middlemen (the casino, the track, the bookie), they're going to take their cut. That leaves the hapless bettor to divvy up the loser's portion. However, when you go one-on-one with a friend, a neighbor, or a business associate, you eliminate the middleman. Of course, you always run the risk of becoming a social outcast if you are too successful. But better you run this risk than succumb to the compulsion to feed the insatiable appetite of the professional gambler. When you go one-on-one, at least you start out with a hypothetical fifty-fifty chance of winning. But if you are smart, you'll do more homework than your friends. You'll know about injuries to key ballplayers. You'll study the recent performance of the combatants, particularly against common opponents. You'll evaluate their coaching staffs, their proclivity for turnovers, their special teams, kicking game, quarterbacks, and which team has the "home court" advantage. But most important, you'll study your fellow bettors. If one happens to be completely freaked out by his favorite team, you can probably strike an advantageous bet. Strong emotional ties, playing favorites, and following hunches are fatal flaws in a gambler.

You go to one and say, "So you like the Cowboys, huh?"

"Best team on God's earth."

"You think so?"

"I *know* so!"

"Yeah? Well, I think the Giants are going to wup them!"

"Are you crazy? No way!"

"The Cowboys will be down after their big win over Pittsburgh and the Giants will sneak up on them and dish it out real good."

"You wanna bet?"

"I'd consider."

"What do you want?"

"I'll take two to one."

"You *are* crazy. You're on. One hundred bucks at two-to-one odds."

"It's a deal. When the Giants win, you pay me two hundred dollars."

"Not that way at all, baby. You have just given me a present of one hundred bucks."

Then you go to a Giant fanatic. "Yech, those Giants are lousy. Can't punch their way out of a paper bag."

"Hey, wait a minute. You're talking about my team."

"You a Giant fan? I thought their fans were as extinct as the dodo bird."

"You got a big mouth."

"Sorry, didn't mean to offend. But even you can see they're going nowhere fast. The Cowboys will swamp them."

"Waddya mean the Giants and 17 points is too much!"

"You wanna bet?"

"I'm not a betting man."

"Put up or shut up."

"If you insist. I'll take the Cowboys, even odds."

"Are you crazy or something? Even odds against the Cowboys? You gotta be crazy."

"Well, the Giants are your team. I thought you believed in them."

"I *do* believe in them, but I'm not crazy like you."

"I'll tell you what I'll do. I'll give you fifteen to twelve odds. The Giants win, I pay you one hundred fifty bucks; you pay me one hundred twenty when the Cowboys clobber them."

"You're on, sucker."

Now let's see what sort of a mess you have got yourself into.

If the Cowboys win—

1. you pay Cowboy fan $100;
2. you collect $120 from Giant fan.
3. Your net profit? Twenty bucks.

If the Giants win—

1. you pay Giant fan $150;
2. you collect $200 from Cowboy fan.
3. Not bad. A profit of 50 smackers.

If there's a tie, all bets are off.

On balance, not a bad mess. Put simply, you can't lose. Besides, it gives you a team to root for. A Giant win will pay off thirty bucks more than a Cowboy romp. Your allegiance to football and the Giants should fill you with a deep sense of patriotic pride. It is very American to worship the tube on Sunday afternoon. It is even more American to root for the underdog. Finally, as serendipity, you're going to make someone happy half the time. And all the time, except when there's a tie, you'll make yourself happy. Golly, you're a good guy/gal.

TECHNICAL NOTES FOR CHAPTER 1:

In chapter 9 we discuss the use of z-scores to assist us in interpreting a test score. Well, a variant of that z-score permits us to estimate the likelihood that in any given day a sample of births of a given size will contain 50% or fewer males. The formula is:

$$z = \frac{p - p_o}{\sqrt{\frac{p(1 - p)}{N}}}$$

in which p = the proportion of white males born in the population ($p = .5141$).

p_o = the hypothetical proportion we are interested in assessing. In the present example, $p_o = .5000$ since we are interested in assessing the probability that the proportion will be as low as .5000 in any given sample.

$1 - p$ = the proportion of females born in the population ($1 - .5141 = .4859$)

N = the number of newborn infants in the sample. Substituting in the formula:

$$z = \frac{.5141 - .5000}{\sqrt{\frac{(.5141)(.4859)}{740}}} = \frac{.0141}{.0184} = 0.77$$

Now we refer to table 9.1. Under $z = 0.7$, we find 24% of the cases; and, under $z = 0.8$, about 21% of the cases. Since $z = 0.77$ is approximately 7/10 of the distance between 24% and 21%, we can say that about 22% of the time, a negative cash flow will be experienced by the gamblers.

CHAPTER **2**

One Servant, Many Masters

Imagine the following scenario. You are bonded in servitude to many masters. They are a motley crew. All are hard taskmasters and will tolerate only a total effort on your part. Moreover, they assume you are available twenty-four hours a day, seven days a week, and fifty-two weeks a year. Your masters span the entire spectrum of physical attractiveness, moral fabric, and social concern. Some are well groomed, considerate, and rational; others wear grubby clothes on grubby bodies topped by grubby minds. You are expected to be loyal and obedient to all—the politician, the gambler, the marketing expert, the advertising executive, the scientist, the minor governmental luminary, the oil company executive, the businessperson, and the head of a large semiconductor industry. Do you wonder what a typical day in your life is like? Let's eavesdrop a moment and see.

"Stat, come here a sec." It's Irma Longwind. She's trying to unseat the incumbent congressperson in her district. Like you, she has untapped reserves of energy. At times she seems to survive on willpower alone. You have never known her to pause for a moment in the heat of a campaign. "Here's what I want you to do," she says. "Find out the voter reaction to my last debate. What do they think of my proposal to lower taxes by abolishing five thousand governmental agencies? Give me a breakdown by political party, ethnic group, gender, and age of the voter. Tell me how I compare to my opponent. Am I making up any ground on him or am I merely treading water? Find out voter reaction to his stand on abortion, gasoline rationing, and teen-age acne. Find out where he is vulnerable. I want all this information first thing last night, understand? When you get that done, I'll really put you to work."

Before you can reply, you receive a peremptory summons

from another master. The gambler. Yech! Such an unsavory character. His life is dominated by one and only one theme—making money, *beaucoup* money, without prompting the growth of calluses on any part of his anatomy, other than his fat derrière. But serve him you must, as you must serve all of your masters. "Stat, get your big A right over to the casino. Now, and I mean right now! A couple of our roulette wheels seem to be coming up with certain numbers more often than they should. They may have developed a flat spot on one of the bearings. But before we strip the friggen things down, we wanna know if they're acting queer."

"Stat, oh, Stat. Before you go, could I talk to you a moment?" It's that lovely gal from marketing. The sweetest one of the bunch. Always so warm and friendly. You wish you had more time to spend with her, but there are only so many hours in the day. "We've reached an impasse on the packaging of a new product. Three different designs have been proposed by our people in the creative department. Each person insists that their design is the best. No one will budge. I proposed that we test

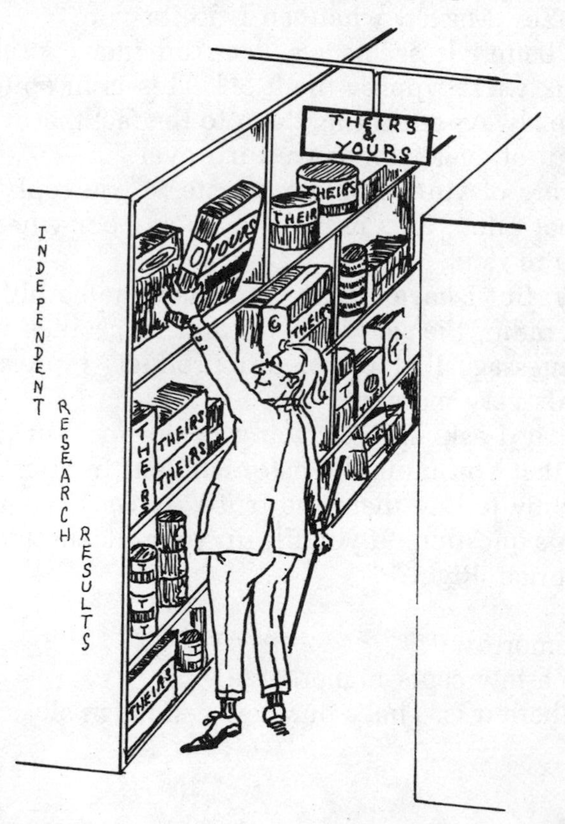

market the product in all three packages. We'll display them in selected markets throughout the U.S. Can you tell us which one will be preferred the most when we go into general distribution?"

"Yes," you reply, rather pompously. "As long as you collect the data according to my specifications. In fact, if you do your job right, I'll go one step further. I'll give you an estimate of the nationwide sales once the packaging issue is decided."

"Oh, Stat, you're a dear. I knew I could count on you. You never seem to let us down."

Before you can savor the triumph of the moment, another voice penetrates the ether. Strident, important, self-indulgent. You know immediately it's that jackass from the oil company. Always trying to prove to the public that he loves them, that he lies awake nights thinking up ways to serve them better, and that he is oh, so friendly. Actually, he is close to being truthful. Just change *love* to *loathe*, *serve* to *screw*, and *friendly* to *hostile*, and you have it.

"Yes, what can I do for you?"

"Yes, Master."

"Sorry. Yes, Master, what can I do for you?"

"That's better. It seems we have an image problem. Some idiots think we're ripping them off. This is in spite of the fact that we are always educating them to the fact that we only make pennies out of every dollar we turn over."

"I'm aware of your problem, Master," you reply, affecting a very somber mien, "It's terrible to be so poor when everybody thinks you're rich."

"Exactly. But I have a plan for a new media blitz. We'll saturate TV, radio, the newspapers, and magazines with the undeniable message that petroleum products represent fantastic buys *at this very moment*."

"How may I ask, are you going to do this?" It is with some difficulty that you mute the astonishment in your voice.

"Simply by telling them the truth, as we have always done. Answer this question: 'If you fill up your guzzler today, you pay a certain price. Right?' "

"Right."

"And tomorrow?"

"It'll be a few cents higher."

"Well, there it is. That's our new campaign slogan. The greatest bargain on earth. Every time you buy it, it's cheaper than the next time. How do you like that? Do you think it will float?"

"Yes, Master, like a cow chip in a punch bowl."

"Er, Mr./Ms. Stat, do you . . . er . . . have a moment?" It's that nervous civil servant. He's trying to turn in a credible performance, but someone's always nosing about to see if his job is expendable. Nervous as a banshee but a nice guy, nevertheless. Not like some of those minor officials who affect an air of great importance in the hope that it will mask the bankruptcy of their mission. "I'm doing a study on the causes of absenteeism in the pajama industry," he says, in a hesitant voice, perhaps anticipating some crude remark. But not you. No way you're gonna ask how much they pull down a week.

"Yes," you say in a kindly manner, since you recognize a kindred spirit—both of you are servants with many masters. "What is your problem?"

"Well, I recently initiated a survey in which I asked the employer to provide the percentages of various types of absentees, broken down by sex."

"So?"

"I just got a reply from one employer. She said that there were three people out with the flu, one with a severe head cold, one with jock itch, but none broken down by sex."

"Aha! Your problem is the use of the word *sex*. It's ambiguous these days. You should substitute the word *gender*. Nobody goes to bed to have gender."

You feel a rush of warmth in your abdomen as you see the light of understanding come to his face. Again, the satisfaction is short-lived. Another strident voice pierces the air.

"Hey, hold on a cotton-pickin' minute. Where do you think you're going?" It's that advertising executive, boisterous in voice, gaudy in dress, and dishonest in everything else. "We just landed a new account. A biggee! I want you to prove that their product is better than the nearest competitor's."

You're tempted to say, "Why, that's easy. Just tell a lie." But you don't. With the emphasis on truth in advertising by umpteen federal agencies these days, you can't be bold-faced about it any more. Get yourself elected to Congress if you want to lie with impunity, but the media are no longer so privileged. Lying must be done with consummate subtlety, finesse, and discretion.

"How about the multi-independent-research-organization scam?" you suggest.

"The one in which we hire a couple of dozen different independent research firms to run exactly the same study and then cite the most favorable one in our advertising?"

"Exactly."

"I think we've overdone that one a bit. Have any other ideas?"

Before you can answer, you are interrupted again, quietly, insistently, urgently: "I need your undivided attention. Can you give it to me?"

"Why, of course." Of all of them, she is your favorite. Her approach is quiet, firm, and sincere. All she asks of you is truth, plain and simple truth. No shenanigans, no messing around with numbers, no use of words to convey ambiguous meanings.

"That last compound we tried seems to be active in the right places. We're ready for a full-scale evaluation of its effects on laboratory animals. We want to set up four experimental groups in which we administer the compound in varying amounts. We want a full blood workup, evaluation of the myelin sheath, spinal fluid—"

"Can do!" you reassure her. You realize that you're not supposed to play favorites. Treat everybody alike, that's your motto. But there are times . . .

"Boy, am I glad I ran into you." It's that nonstop member of the industrial jet set. His company manufactures microcomponents for the electronics industry. "We're having quality-control trouble. It seems that no matter what we do, defective components keep getting shipped out. I'm getting tired of hearing complaints wherever I touch down—Ankara, Rome, Paris, Tokyo, Sri Lanka, or Mamma Luigi's Spaghetti Factory. Complaints, complaints. I'm tired of hearing complaints! Understand? I want you to run down to the plant and see what's going on. If we don't get a handle on this thing, we're out of business."

Well, there it is in a nutshell, and I haven't even told you half of the story. There is a steady stream of mendicants who pass by your door daily, hoping for a bit of wisdom that will prove them right. You have a heavy burden on your shoulders, and it's not really appreciated. It's not bad enough that you have so many masters. To make matters worse, there are those millions of rowdy spectators whose greatest pleasure in life is to sit on the sidelines and poke fun at you. You've heard what Disraeli said, haven't you? "There are lies, damn lies, and statistics." It has even been said that a politician uses statistics in much the same way that a drunk uses a lamppost—more for support than illumination. Others have likened statistics to a hooker—available, compliant, amoral. Need we be reminded of the fact that without customers the profession would have zero membership? It seems that like the unfortunate oil companies, statistics has an image problem. It is known by the company it keeps. In fact, even its birth has a cloud over it. It was conceived and delivered in the gaming houses of France. Its sire, appropriately enough, was a wealthy playboy known as Chevalier de Méré.* It seems that the family fortune was not sufficient to support him in the life style to which he aspired. He worked out an ingenious scheme to supplement the family coffers. It involved, perhaps, the first sucker bet. It goes like this: "I'll take even odds that if I roll a die four times, I will obtain at least one 6." He reasoned as follows: "If I toss it once, the chances of obtaining a 6 are one in six; twice, it's two in six; three times, the chances are three in six, otherwise known as even odds. If I toss it four times, the chances are four in six. That's two to one in my favor. A can't-miss proposition." As expected, he cleaned up. But he was right for the wrong reasons. Had he kept a careful record of his winnings, he would have discovered that the ratio of wins to losses was about 1.08:1 rather than the 2:1 he had theorized. In fact, had he carried his reasoning two steps further, he surely would have discovered an error in his reasoning. "If I toss the die five times, the chances of obtaining one 6 is five out of six. That's five-to-one odds in my favor. Six tosses will yield a ratio of six out of six. Why that's a certainty! There absolutely *must* be at least one 6." Oh, yeah? Anyone familiar with the behavior of galloping dominoes would recognize immediately that certainty is not one of its enduring traits. For the record, the chances of obtaining at least one 6 in six tosses are only about 2:1. That's far from certainty in my book.

Nevertheless, de Méré's bet was a good one. He found his coterie of suckers and the moola rolled in. Well, if one scam

* His real name was Antoine Gombauld.

is so successful, there must be others. One thing he knew at this point: There were plenty of pigeons waiting and wanting to be plucked. He would oblige them. He began a simple variation of his previous success, using two dice instead of one. His reasoning was the same. "I'll take bets, even odds, that I will obtain at least one 12 in twenty-four tosses of two dice." His prior triumph left no room for doubt that his new venture would be equally successful. "In twenty-four tosses of a pair of dice," he reasoned, "the chances are twenty-four in thirty-six or two in three of getting at least one 12. Those odds are still two to one in my favor." But suddenly, unexpectedly, the plucker was plucked. Unbeknownst to him, the odds had now shifted against him. Not very much—indeed, a very small amount. Nevertheless, as small as it was, he was suddenly paying it out more often than he was raking it in. Shaken, stunned, appalled, and with his tail between his legs, he approached Blaise Pascal, the famous French mathematician. "What went wrong?" he pleaded in anguish.

Pascal and Pierre de Fermat, a fellow mathematician, put their heads together and worked out the complete set of probabilities for this game of chance as well as many others. Probability and statistics had taken a first halting step in the affairs of humanity, and the world has never since been the same. Insurance companies arose, like the famous Lloyd's of London. Initially no more than the riskiest sort of speculative ventures, dealing mostly with the shipment of goods over hazardous waters, they have since taken on respectability. Even today, however, they are little more than thinly veiled but highly sophisticated gambling operations. Life insurance companies gamble on your staying alive sufficiently long that they'll make a bigger profit from the use of your money while you're alive than they will ultimately be forced to pay out. An occasional spoilsport will beat the odds by popping off early. But the victory is purely Pyrrhic, let me assure you.

But let's not judge statistics by the company it keeps. In the coming pages, we shall glimpse a broad panorama of things done in the name of statistics. We'll look at the seamy side in the hope of entertaining while exposing statistical fraud. We'll also look at its good face, the one that caused H. G. Wells to predict that the day would come when the ability to think statistically would be as important as the ability to read and write.

TECHNICAL NOTES ON CHAPTER 2

Page 27: It is relatively easy to calculate the probability of obtaining at least one 6 (i.e., one or more 6s as long as we restrict ourselves to this limited case). It is 1.00 − $(5/6)^N$, when N is the number of tosses and $(5/6)^N$ directs us to raise the fraction to the Nth power.

Thus for:	**Probability of at Least One 6**	**Odds in Favor**
1 toss (N = 1): $1-(5/6)^1 = 1-5/6$	$= 1-.83 = .17$	1:5
2 tosses (N = 2): $1-(5/6)^2 = 1-25/36$	$= 1-.69 = .31$	1:2.23
3 tosses (N = 3): $1-(5/6)^3 = 1-125/216$	$= 1-.58 = .42$	1:1.38
4 tosses (N = 4): $1-(5/6)^4 = 1-625/1296$	$= 1-.48 = .52$	1.08:1
5 tosses (N = 5): $1-(5/6)^5 = 1-3125/7776$	$= 1-.40 = .60$	1.5:1
6 tosses (N = 6): $1-(5/6)^6 = 1-15625/46656$	$= 1-.33 = .67$	2:1

Note that odds are calculated in the following way:

1 toss: .17/.83 = 1:5 (rounded)
2 tosses: .31/.69 = 1:2.23
3 tosses: .42/.58 = 1:1.38, etc.

The probabilities (and odds) shift in favor of obtaining a 6 when there are 4 tosses of a die. However, rather than being 2:1 as Chevalier de Méré thought, the odds are only 1.08:1. Only when a die is tossed 6 times are the odds favoring a 6 as high as 2:1. This is far from the certainty de Méré may have thought.

The chances of getting at least one 12 in 24 tosses of a pair of dice is:

$$1 - (35/36)^{24} = 1 - 0.5086 = 0.4914$$

Thus, the probabilities turned ever so slightly against de Méré. Thus, he went from a winning position to that of a loser when he changed the "game."

Incidentally, the general formula is:

$$(P + Q)^N = P^N + {}^{N}\!/_{1}P^{N-1}Q + \frac{N(N-1)}{(1)(2)}P^{N-2}Q^2 + \ldots Q^N,$$

in which P is the probability of obtaining a 6 on one toss of a die (i.e. $P = 1/6$) and Q is the probability of obtaining a non-six (i.e. $Q = 5/6$). The probability calculation is simplified by the fact that the probability in which we are interested (one or more sixes) includes all values except zero sixes. Consequently, we may use a simplified formula in this special case, viz., $1 - Q^N$.

CHAPTER 3

The Two Faces of Statistics

If you're going for a swim, you might just as well jump right into the water and get over that initial chill all at once. If you're going to be a statistician, the same principle applies. So we are going to make you a statistical consultant to a parts supplier for the aerospace industry. Here's your problem. The company manufactures hundreds of thousands of items a day. Some are so precise and so small that a microscope is a required item of inspection equipment. At this very moment, your company has a contract to produce millions of valves to control the flow of gases in a combustion engine. The contract calls for these valves to be ten millimeters in diameter with an error of no more than .1 millimeter. If more than one percent of a day's output exceeds these tolerance limits, a severe financial penalty is imposed that could, if repeated often enough, bankrupt the company. So the stakes are high Mr., Mrs., Ms. or what-have-you Expert. What do you do?

What's that? First thing you'd do is figure out how to measure those little critters? Good start. The measurement problem is the first one you must solve. If your data base—your raw measurements—is not accurate, you might just as well scrap the whole effort. That's where public opinion polls and surveys encounter their greatest difficulties. You can use the most advanced methods to select the sample and analyze the results with incredibly sophisticated high-speed computers, but if a number of your respondents chose not to answer honestly, your results are as useless as last night's theater tickets.

Why would someone not answer honestly? Status and self-image, for starters. If the answer poses a threat to the individual's self-concept or public image, he or she might well falsify the answer. "Do you subscribe to *Hustler* and/or *Penthouse*?"

"Why, of course not. Just what do you think I am?"

"Well, what magazines do you regularly receive?"

"*National Geographic, Wildlife,* and *Arizona Highways*. Oh,

"Are you in favor of the government building dams for flood control?"

yes, and *Saturday Review*. I read it cover to cover every Saturday."

"*Saturday Review*'s not a weekly. It hasn't been for years."

"Oh?"

You may also encounter some difficulty if you ask an ambiguous or leading question.

Let me illustrate. Back in the 1930s, farmers along the Ohio River were being wiped out year after year by the spring floods. It was proposed that the government build a series of dams for the purpose of flood control. When the farmers of annual grief were asked, "Are you in favor of the government building dams for flood control?" the response was overwhelmingly favorable. But when the question was slightly reworded: "Are you in favor of the socialistic practice of the government building dams for flood control?" the farmers voiced a resounding "Nay!" So it is that we often respond to words rather than issues, claims rather than facts.

"Are you in favor of the socialistic practice of the government building dams for flood control?"

OK. Let us assume that you have found a foolproof way to differentiate the bad valves from the good ones. What do you do next? Find some way to select representative samples of the day's output? Good. Couldn't have stated it much better myself. But why not test every valve?

Now calm down. Please don't get so excited. I wasn't *telling* you to test the lot. I am only asking, Why not?

Good. You're absolutely right. There is no way that you could economically test the entire day's production. The company would, as you say, go down the tube in no time at all. That's not exactly the way you phrased it? But I think I caught your meaning without offending anyone. So what's your alternative? Random sampling procedures? Good. Excellent. And how would you do that? I understand. You are the boss statistician and you concern yourself only with broad strategies and avoid the everyday nitty-gritty details, right? So you would hire a statistical consultant? I'll buy that. Actually, I didn't want to go into random sampling at this point either. Suffice it to say that it is merely a method to ensure that each sample of a given number of valves is equally likely to be selected from what we experts call a population. In this case, the population is the entire day's output.

Now that you have collected a number of samples of valves, what would you do next? You would determine the proportion of defectives in each sample? Excellent. This takes us into the first function of statistical analysis—the *descriptive function*. After you have collected a mass of data, you want to manipulate the raw facts in ways that permit you to make summary statements. If you collect 100 valves in a sample and find that 3 of them are defective, you could summarize the results for the sample by saying that the proportion of defective valves is 0.03, or 3 percent. For other types of data, you may be interested in using other descriptive statistics, such as the arithmetic average or the range.

But your work doesn't stop there, does it? It's rare that we are interested in the descriptive statistics as such. Knowing that there are 3 defective valves in a sample of 100, or 6 out of 15 families watching "Barney Miller" does not send us into paroxysms of joy. What does light a fire under our flagging interest?

Precisely, our estimate of the proportion of defectives in the population (the whole day's output) of valves. If the sample proportions lead you to estimate a proportion of defectives greater than 1 percent, the company is in trouble. The engineers will have to go over the manufacturing process in the greatest detail to correct the trouble. On the other hand, if it appears safe to conclude that the proportion of defectives is less than 1 percent, we are in clover. We can continue to satisfy our customers while raking in the moolah. In any event, whenever we use the statistics taken from a single sample or from a number of samples to estimate characteristics of the population, we are taking a giant inductive leap into the unknown from the known. This leap reveals the second face of statistical analysis—the *inferential function*—in all its glory.

In many ways, inferential statistics is the most exciting aspect of the statistical ball game. Whenever you take an inductive leap, there is always an element of risk involved. Pollsters sometimes call the wrong winner, as Clem McCarthy did some years ago when calling the Kentucky Derby. Occasionally a drug is released that is not really "safe" or "effective." The beauty of statistical inference is that while admittedly involving the risk

Box 3.1 Confusing a Datum with a Statistic

How often do you hear the M.C. at a beauty pageant announce, "Her vital statistics are 36, 23, 34"? These aren't statistics. They're data. A statistic is a statement that summarizes a collection of measurements. It would be correct to say that the mean (a statistic often referred to as the average) vital statistics of all contestants are 36.4, 23.6, 34.3.

Similar confusion arises whenever a commercial makes such statements as "Come to Junk-heap TV Repair. Our service personnel have one hundred years of repair experience," or "In more than a million grueling miles of testing, our automobiles had a repair record of less than one percent." What the announcers carefully left unsaid were the number of repairmen in the TV shop and the distribution of their work experience and the number of automobiles tested. For all we know, the boss may have had twenty-five years of experience, with each of seventy-five underlings able to claim only one. Also I would not be terribly impressed to learn that 100,000 new cars had been put through ten miles of testing each. How many breakdowns are likely to occur during the first ten miles in the life of a new car?

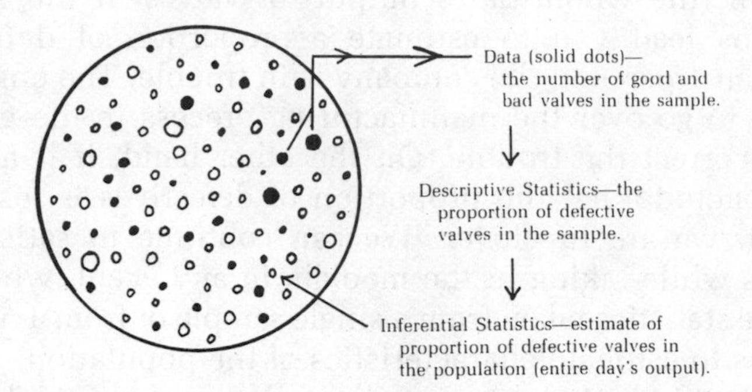

Fig. 3.1 The Inductive Leap. A random sample is selected from a population of interest. The dots represent all valves produced in a day (the population). The solid dots represent the valves in the sample. The data are analyzed in terms of the proportion of defective valves in the sample. This proportion is the descriptive statistic. In inferential statistics, we estimate the proportion of defective valves in the population (i.e., the entire day's output). [Adapted from R. P. Runyon and Audrey Haber, *Fundamentals of Behavioral Statistics*, 4th ed. (Reading, Mass.: Addison-Wesley Publishing Co., 1980).]

of an incorrect conclusion, it permits us to estimate precisely what the risk is. And we can raise and lower the risk, as the situation demands. But "I'm getting ahead of myself. We'll be returning to this fascinating aspect of statistics on and off throughout this book. Take a look at figure 3.1 if you want a summary of the steps involved in going from the collection of data through estimating population values.

Those Mischievous Media Merchants

Confusion between the two functions of statistical analysis can be a great source of mischief. The TV pitchman solemnly intones, "Sixty percent of doctors interviewed prescribe WIB for the relief of tension headache." This "60 percent" is a descriptive statistic. It could be based on as few as five interviews (if three say they prescribe WIB, that's 60 percent). But the trouble is that most people remember the statement that 60 percent of doctors—the whole population of doctors rather than the sample—prescribe WIB. They have made the inductive leap without realizing it. And the words of doctors carry much weight in our society.

Then there is the other form of hocus-pocus that raises the non sequitur to the level of a fine art. Here's what you do. You cite research results that have nothing to do with the advertised use of the product but—pay attention, this is important—in so doing, you must keep an absolutely straight face. Otherwise, the audience will discover the deception. My nominee for the all-time champion of the non sequitur is the late David Janssen, who solemnly informed us that studies made by a leading university showed that Excedrin is more effective for "*pain other than headache*" (italics mine). He then concluded with cherubic innocence and sincerity, "So the next time you get a headache, try Excedrin." How many of you caught that non sequitur? Of course if he had said, "So the next time you get cancer, try Excedrin," we would probably have noted the transgression. Oh, if the participants weren't so damned serious and self-righteous about it, TV lies could be fun.

Or maybe the statement is based on a larger and more representative sample. Let's look at this beautiful little bit of chicanery: "Of three thousand doctors interviewed, eighty percent prescribe WIB for the relief of tension headache." That's a lot of doctors, most impressive! But let's look at the interview.

"Doctor, do you prescribe OAT for the relief of tension headaches?"

"For some patients, yes."

"COH?"

"Yes, for some patients."

"IRD?"

"On occasion."

"WIB?"

"Yes, on occasion."

The doctor prescribes WIB so it goes down as a "yes" response. But nobody tells us that he or she may also prescribe everything else under the sun, including snake oil.

Or, how about this one: "Ninety-five percent of the doctors prescribe the pain reliever found in WIB." What is left unsaid is the fact that the same pain reliever appears in OAT, COH, and IRD. This is very similar to the claim "Drivers taking the first three places at Indy wore Bliby's Bloomers." What we are not told is that every driver in the race wore Bliby's Bloomers. (They also gargled with STP.)

CHAPTER 4

Graphic Gullduggery

Don't look up *gullduggery* in the dictionary. You won't find it. It is a neologism that I coined specifically for the occasion. In the event you haven't guessed, gullduggery combines the meanings of the verb *gull* ("to cheat," "trick," "dupe") and the noun *skullduggery* ("mean trickery," "craftiness"). Why did I coin it? Well, for one thing, I like alliteration. More importantly, graphs are often used to trick or dupe the reader in a mean and crafty manner. There is something absolutely hypnotic about a well-conceived and well-executed graph. It demands your attention while anesthetizing your critical faculties. Take those animated graphs they display on the tube. They really blow your mind. Lines move here and there as if they have lives of their own, bubbles bubble through hollow tubes in the body, hammers labor diligently in the throbbing brain, and red swellings explode in inflamed joints. In the meantime, the voice-over lugubriously intones THE MESSAGE: Medical research has proved that the active ingredient in IRD is the most effective pain reliever that can be purchased without a doctor's prescription. And the caption that, I guess, warns us to take the claim with a grain of salt flashes by so fast it would take a recent graduate of the Evelyn Wood Institute to decipher it. I say guess because they never gave me sufficient time to read the whole damn thing.

Now don't get me wrong. Graphics are the lifeblood of the descriptive aspect of statistics. Just as a picture is worth a thousand words, so is a well-constructed graph worth a thousand data points. There is a veritable wealth of information that can be conveyed with a few lines, a bar or two, or a circle divided up like a pie (cut the pie in half if you want to, but give me the bigger half).

Unfortunately, the very virtues of graphics are their Achilles' heel. They can be attractive, particularly when executed by an artist with a flare for the unusual. However, there is always

Fig. 4.1 Validating Pseudograph. You've seen this one on TV hundreds of times, and so have I. A person is providing expert testimony on the efficiency of some product that he or she is hawking on TV. In a graph shown to the side, the lines move in such a way that they appear to prove the validity of the testimony. Great care is taken never to label the axes. For example, fig. 4.1 discusses a hypothetical engine additive that promises to add new life to the geriatric set of gas guzzlers.

someone poised on the sidelines waiting to peddle the pulchritude in the bawdyhouses of commerce. Like the skilled stage magician, the artist can direct your attention to one part of the graph in order to distract you from another part, where monkey business is going on. Perhaps the best example of graphic legerdemain is in the type of deception I call the *validating pseudograph* (see figure 4.1).

Actually, this type of deception doesn't bother me too much. It is so obviously dishonest that I can sit back and enjoy its ineptitude, in the same way that I occasionally enjoy a grade-B movie or the usual TV fare.

But *rubber band boundaries* are something else again.

The Rubber Band Boundary

What makes the *rubber-band-boundary chart* particularly insidious is that it is generally used to present important data and

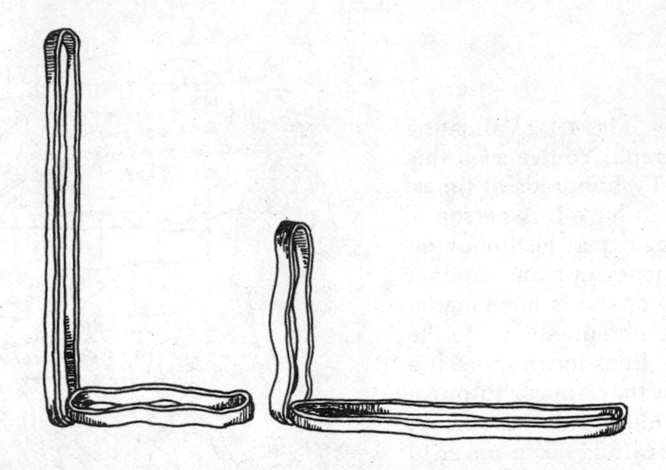

commonly appears in serious publications (I have seen them repeatedly in the most highly respected newspapers as well as in weekly news magazines). The rubber-band-boundary chart is really the progeny of graphic anarchy. By this I mean that there are no universally agreed-upon methods of representing the relative lengths of the vertical and horizontal axes. Therefore, these axes are like rubber bands, ready to expand or contract on demand of the user.

Let us say that the user is a sales manager of a department in a large discount store. He has kept a careful record of the sales figures of all the personnel working under him, and he wants to use these figures to put a burr under the saddle of salespeople who, he feels, are not giving their all to the dear old shop. What does he do? He draws bar graphs representing the weekly dollar volume for each salesperson, but he stretches the vertical axis and contracts the horizontal axis. He states, "Ms. Dee, you are obviously not holding up your end of the workload. The rest of the department is carrying you. If you don't believe me, look at the sales chart." (See figure 4.2a.)

But Ms. Dee has not been caught off guard. Without her manager knowing it, she has been taking prolonged coffee breaks each afternoon while preparing the hors d'oeuvre shown in figure 4.2b. She replies, "It's strange to hear you say that. As you can see, I'm doing about as well as anybody. A few sales, one way or the other, and I would be the top in the department."

It is because of the elastic axes that I have long advocated the

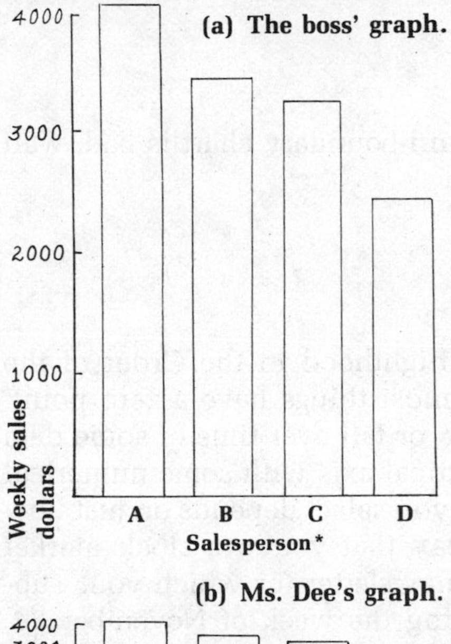

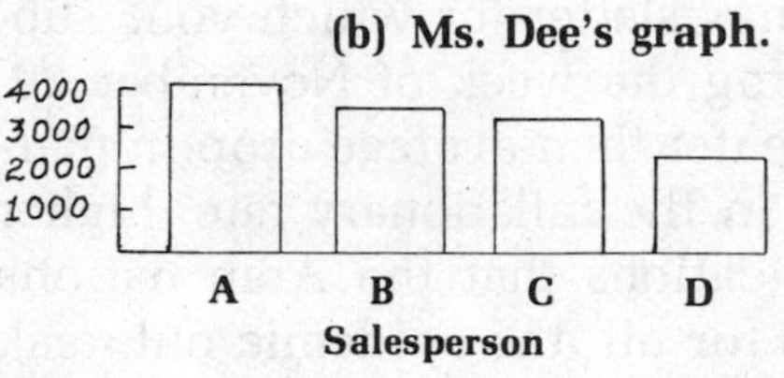

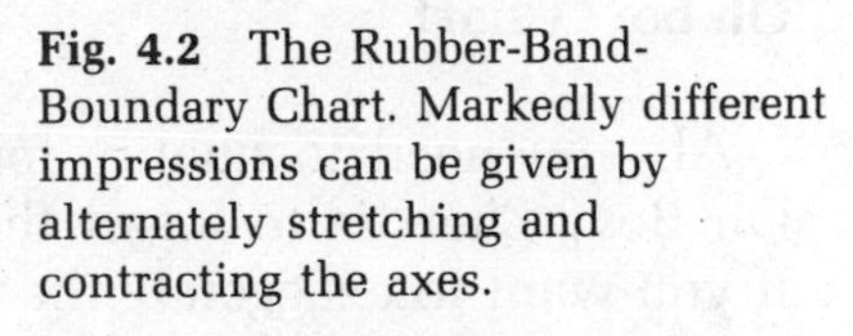

Fig. 4.2 The Rubber-Band-Boundary Chart. Markedly different impressions can be given by alternately stretching and contracting the axes.

*Ms. Dee tried unsuccessfully to convince the boss to abandon his sexist means of identifying the sales force. He says he'll be damned if he'll participate in the androgynization of the mother tongue.

three-quarter-high rule as a means of ending graphic anarchy. This rule is expressed as follows: The vertical axis should be laid out so that the height of the maximum point is approximately equal to three quarters of the length of the horizontal axis.[1] The rule has the virtue of removing personal preferences and biases from the graphic decision-making process. Figure 4.3 shows the sales data of the previous example presented in accordance with the three-quarter-high rule.

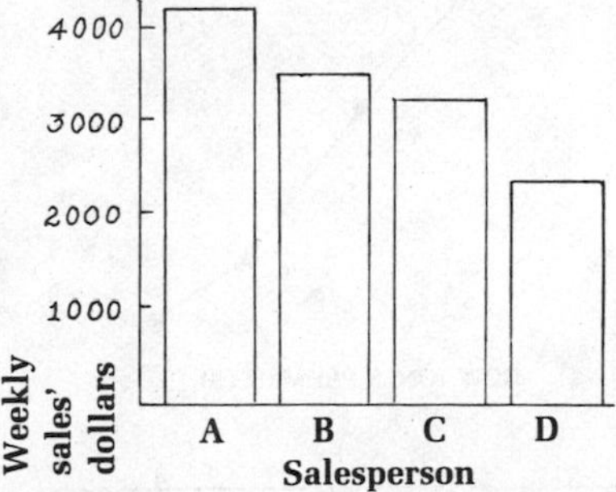

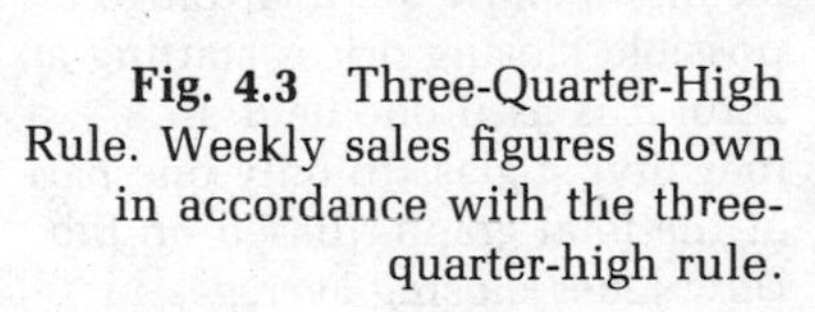

Fig. 4.3 Three-Quarter-High Rule. Weekly sales figures shown in accordance with the three-quarter-high rule.

[1] R. P. Runyon and A. Haber, *Fundamentals of Behavioral Statistics*, 4th ed. (Reading, Mass.: Addison-Wesley Publishing Co., 1980).

But if you think the rubber-band-boundary chart is bad, wait until you see the *Oh boy! chart.*

Oh boy! Chart

All you need to qualify for knighthood in the Order of the Oh Boy! Chart is to forget that most things have a zero point. If you want to exaggerate the rise or fall over time of some data points, you merely begin the vertical axis with some numerical value other than zero. The value you select depends on just how unscrupulous you are. Let us say that you're a stock market analyst who puts out a weekly newsletter for which your subscribers pay a pretty price. During the week of November 24, 1975 you warned, "Look for a greater than-average drop in market prices because of instability in the inflationary rate, higher prime rates, and continued indications that the Arab nations will be looking for higher prices for oil, the epidemic outbreak of civil strife throughout the world, and economic indicators that show we are not emerging from the recession as fast as we would like." You then show the chart of figure 4.4 as proof positive of your prophetic abilities.

Your competitor also puts out a newsletter. In it, the following prediction was made: "Look for relatively stable or moderately

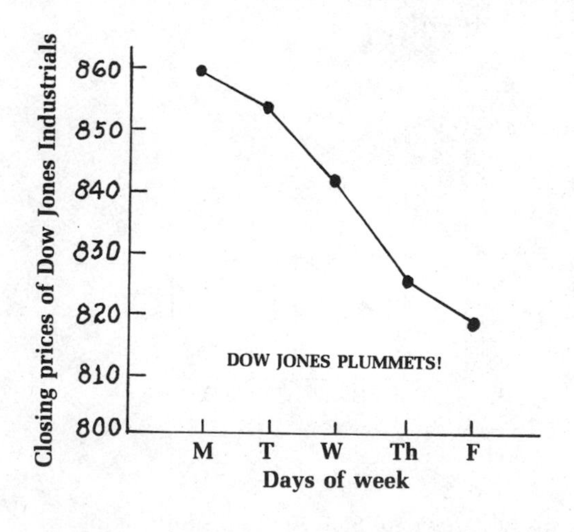

Fig. 4.4 Oh boy! Chart. The falling trend is exaggerated by failing to show the total range of possible closing prices starting at zero. It is as if one held a magnifying glass to only one part of the total graph. (Based on the Dow Jones closing averages in the New York Stock Exchange during the week of December 1, 1975.

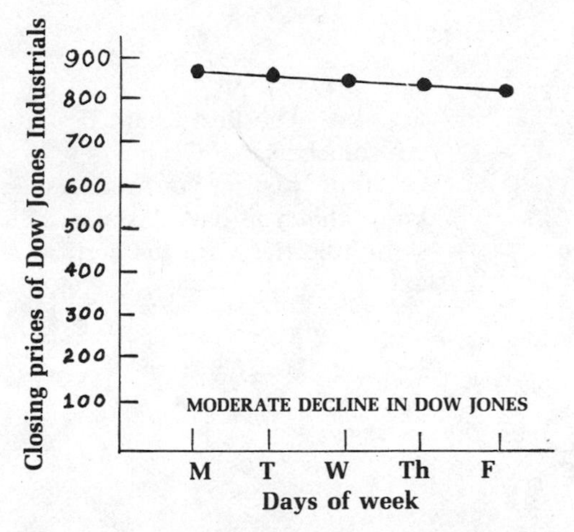

Fig. 4.5 Competitor's Chart. Graph of closing Dow Jones Industrials during the week of December 1, 1975. Note that the full range of possible declines is shown.

falling market prices because of mixed indicators; on the one hand, there is some hope that the federal government will get control over double-digit inflation, and continued indications that the Arab nations will be taking a less extreme view concerning oil-pricing policies. On the other hand, inflation continues, and various economic indicators show we are not emerging from the recession as fast as we would like." As proof of this astute economic forecasting, your competitor produces the chart of figure 4.5.

What's that? You are offended by the second chart? Why?

"Well," you say, "by showing the whole schmeer of possible values from zero to eight hundred seventy, you are obscuring a drop that actually took place. Using your method, even a drop of one hundred points would show up as only a minor perturbation. Believe me, baby, a drop of one hundred points in a week would send a bunch of small investors scurrying pell-mell for the gas pipe."

You've got a point there, I must admit. But if you're going to direct attention to only a limited part of the total range, the least you can do is show a distinct separation at the bottom of the vertical axis so people will know what you have done. And don't label the chart "Dow Jones Plummets." Credit your readers with sufficient intelligence to judge for themselves. (See figure 4.6.)

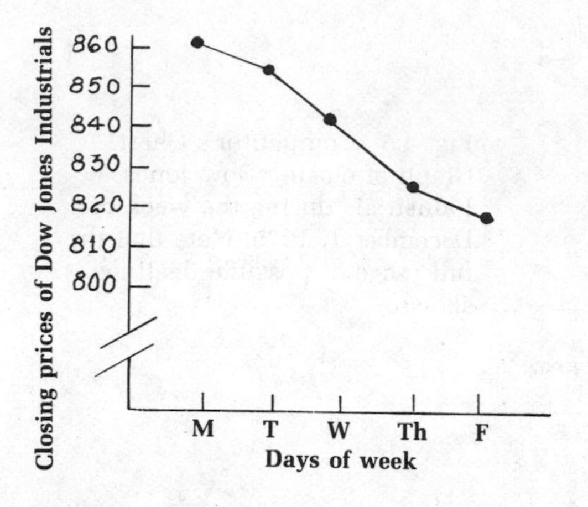

Fig. 4.6 The Best Chart. If you must use an Oh boy! chart, at least let your readers know that you have taken some liberties with the vertical axis.

Popeyed Pictographs

Some people find bar graphs and line graphs exceedingly static, unimaginative, and boring. Instead of bars, they substitute pictures of the things represented in the graphs. Such a figure is referred to as a *pictograph* or a *pictogram*. The Bureau of the Census is fond of using human figures to represent population statistics (see figure 4.7). (They might be accused of some insensitivity, since the last figure to the right is always seriously maimed, lacking, as he or she does, various parts of the body.)

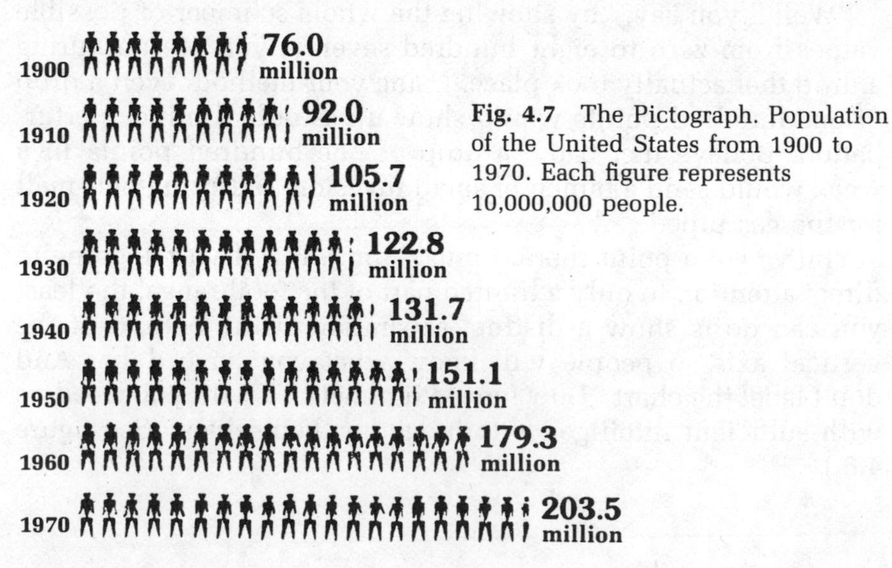

Fig. 4.7 The Pictograph. Population of the United States from 1900 to 1970. Each figure represents 10,000,000 people.

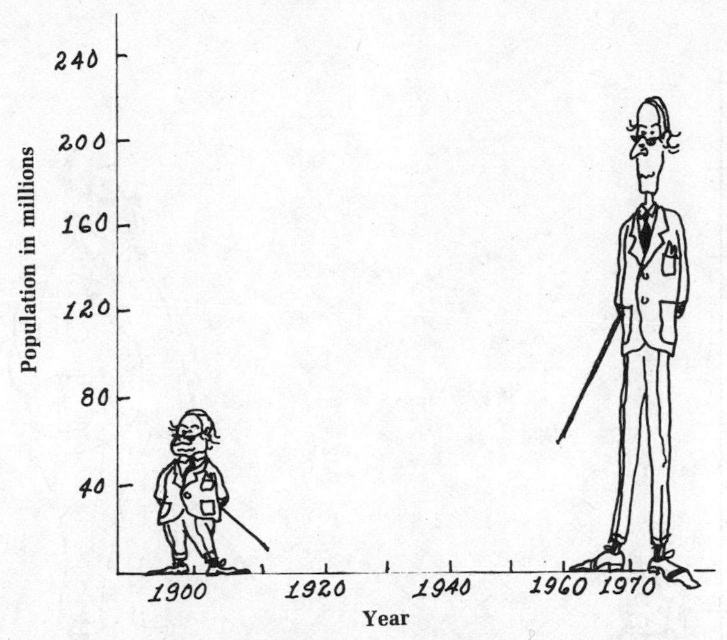

Fig. 4.8 Another version. Pictograph showing the population of the United States from 1900 to 1970.

Pictographs are perfectly legitimate ways of displaying statistical facts. They can be both interesting and, if done correctly, informative. But there's always someone trying to improve on Mother Nature. So, instead of using hundreds of tiny little human figures, our budding genius reasons, "I'll draw just one figure, but I'll make it proportional to the total. Serendipitously, I'll also avoid the dissected human figure." The product of his cerebrations is displayed in figure 4.8.

Our friend ponders his artistic production. "Something seems wrong," he observes. "Why, of course. The human figures are distorted all out of their usual proportions. That poor creature on the right must have spent his last three weeks on the rack. I must restore the proper proportions to the human anatomy." He draws figure 4.9 and is very happy this time with his artistic endeavors. "Not only is it more pleasing to the eye," he observes, "but it somehow makes the growth in population seem even more substantial." Well, of course it does. He has not only improved the proportion of the human figure, but he has vastly increased the total volume.

Granted that the height of the figure is supposed to convey the information about the population, it is difficult to ignore the overall King Kong proportions of the figure to the right. Such

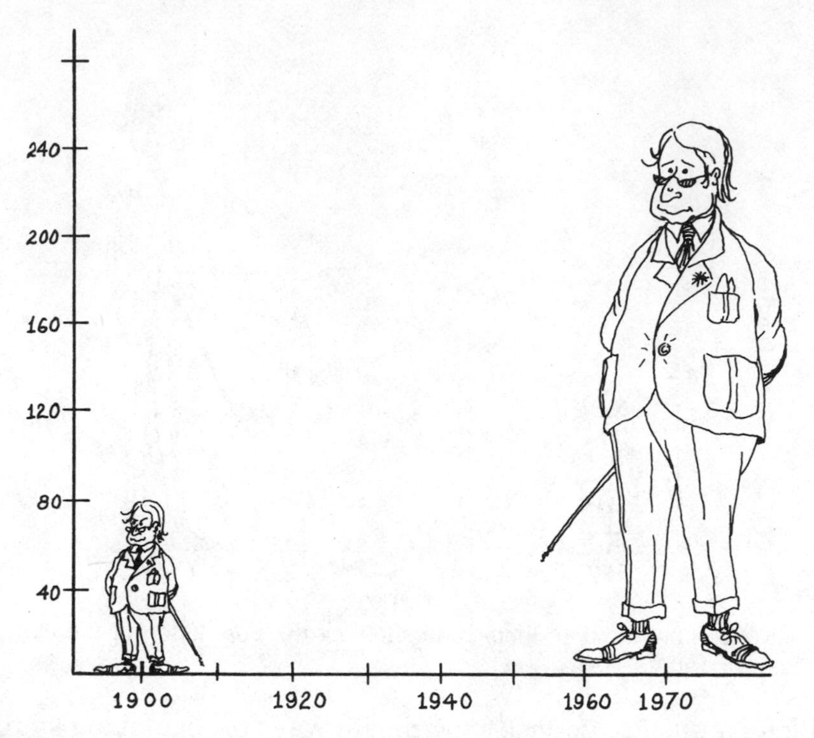

Fig. 4.9 Popeyed pictograph. This shows the population of the United States from 1900 to 1970. Although the *height* of the figure is meant to represent the population, most individuals respond to the *total volume* of the figure. Thus the difference in growth is grossly exaggerated.

figures defeat the basic purposes of honest graphic representation—to convey information interestingly, rapidly, and accurately.

The Double-Whammy Graph

The following type of graph is sheer delight for the perennial prophets of doom. Take a good look at figure 4.10. It tells a twofold tale of grief and despair. First, there has been a steady erosion of the purchasing power of the dollar since 1950. "A dollar today just ain't worth a dollar any more," laments the modern-day Jeremiah (incidentally, I wonder when a dollar was really worth a dollar). But if the decline of the dollar isn't enough grief to heap on the hapless consumer, a rising curve is added to depict the inflationary spiral. The story is clear, albeit dismal. Two equally bad things are happening simultaneously. We are spending a lot more money to buy things because money is less valuable, and at the same time the cost of that new pair of shoes keeps going up.

But wait a minute. There's something fishy here. Aren't these two measures—cost of living and purchasing power of the dollar—two aspects of the same thing, like two sides of the same coin? Of course they are. If you do not believe me, try dividing the purchasing power of the dollar (expressed as a proportion of a dollar, e.g., .511) for any given year, into 100. You should obtain roughly the cost of living index for the same year. Or divide the cost of living (expressed as a proportion of a dollar, e.g., 1.954) into 100, and you will find the value of the dollar for that year. Inflation is bad enough. Let's not compound our miseries by displaying the double-whammy graph.

Fig. 4.10 The Double-Whammy Graph. Double, double, toil, and trouble. As if inflation weren't bad enough, a second factor is added that tells the same story. The result is the appearance of two bundles of bad news. (Source: Data obtained from the *Information Please* Almanac for 1980.)

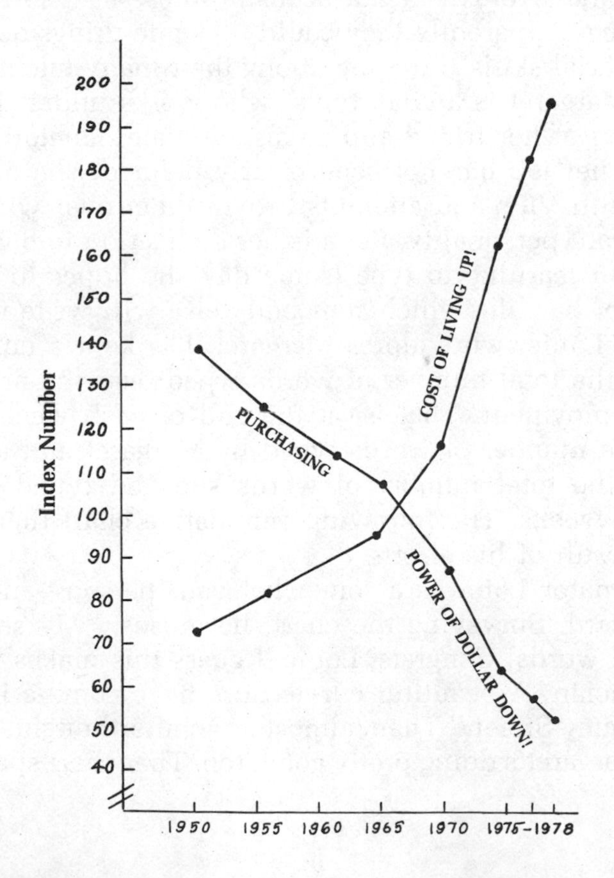

The Cumulative Chart—Only One Way to Go

In this chapter, we have looked at a lot of graphic skullduggery. We have seen axes pinched and squeezed like a baby's backside and stretched like a forty-five bra. We have seen line drawings masquerading as graphs and solid figures blown up like inflatable toys. Now we are going to put the icing on the cake by looking at a perfectly legitimate, extremely useful form of graphic representation that is frequently misinterpreted by both laypeople and scientists alike. I am referring to the cumulative chart, a favorite graphic device of many behavioral scientists.

The cumulative chart is essentially a time/performance graph in which the performance measure is cumulated over time. Do you recall how Congress entertained us a few years ago by hiring secretaries, at the public's expense, who were unable to type? Don't get me wrong. I'm not accusing these secretaries of incompetence. Apparently they could do some things quite well. But secretarial skills were not among their many talents. Let us say that Margaret is a clerk-typist who works under the direct supervision of her friend and counselor state Senator Frumph. Although her job has not been clearly defined, she meets frequently with VIPs and attempts to win them over with charm and a vibrant personality. Such is her dedication to her job that she is even learning to type (some day she hopes to type the memoirs of her life, which somebody else will write for her).

Senator Louie, who adores Margaret, has kept a cumulative record of the total number of words typed over the first weeks of her employment. That is, at the end of each week, he calculates the number of words typed by Margaret and adds this figure to the total number of words she had typed over the preceding weeks. The following cumulative chart (figure 4.11) was the result of his efforts.

Now Senator Louie is a somewhat vain person—an occupational hazard. Surveying the chart, he muses, "My secretary's first typed words! Congrats, Louie. I guess this makes you legit as a politician. A beautiful curve. Sure have come a long way since Delancy Street." Then, almost as an afterthought, he looks again: "Margaret's doing pretty good, too. That chart spells prog-

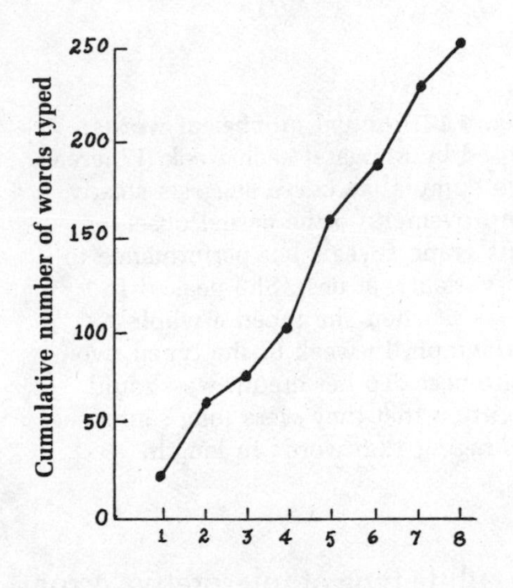

Fig. 4.11 Cumulative chart of words typed per week by Senator Frumph's secretary.

ress, no matter how you look at it. She's getting steadily better. No telling where she will end up."

Louie's mistake in reading this chart is understandable. Like most of us, he is accustomed to interpreting a rising line on a graph as good and salubrious and a falling line as unfavorable at best, and catastrophic at worst. There's only one thing wrong with interpreting cumulative curves in this way. *As long as there are any gains whatever* (number of words typed in Margaret's case) *the line must go up. It cannot go down.* At the very worst, it will parallel the horizontal axis when her typing output drops to zero during any given period.

"We hit a peak today—she typed a whole paragraph!"

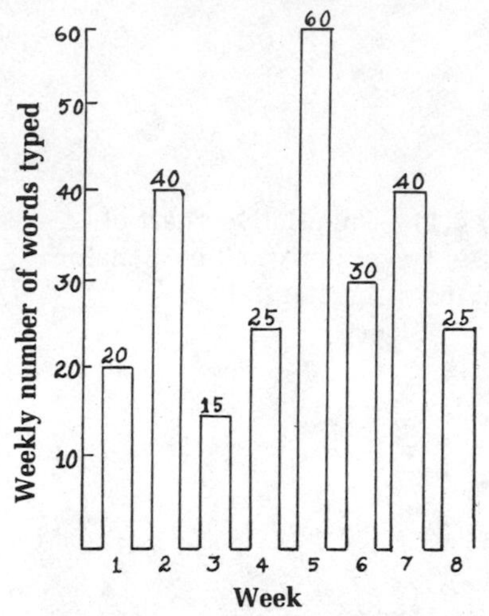

Fig. 4.12 Actual number of words typed by Margaret each week. Whereas the cumulative curve suggests steady improvement to the casual observer, this graph reveals her performance to be variable, at best. She peaked in week 5 when she typed a whole paragraph. In week 8, she typed two sentences. To her credit, we should mention that they were long sentences, averaging 12.5 words in length.

But there is an infinitely more subtle type of interpretive error that cumulative graphs encourage. Since they appear rather smooth and climb steadily from left to right, they seem to reflect stable performance. When a fellow state senator in an adjoining district perused the graph, he commented, "Jeez. That gal is a steady performer all right. A good dependable performer. I could use one hundred like her for my committee work."

Unfortunately, it is in the nature of cumulative curves to appear stable. The cumulative adding of all prior performance tends to smooth out even large differences in performance from one time period to another. In fact, the cumulative curve we have been examining was based on the week-to-week performance figures shown in figure 4.12.

Incidentally, in the event that you encounter a cumulative curve and would like to derive the original time-by-time performance for it, the procedures are very simple (see figure 4.13).

1. Measure the height of the curve for the first time period. Subtract this amount from the curve representing the second time period. The difference is the performance during the second time period.

2. Measure the height from the horizontal axis to the rising curve for the second time period. Subtract this amount from the curve representing the third time period. The difference is the performance during the third time period.

3. Repeat these procedures until the original performance at each time period has been derived.

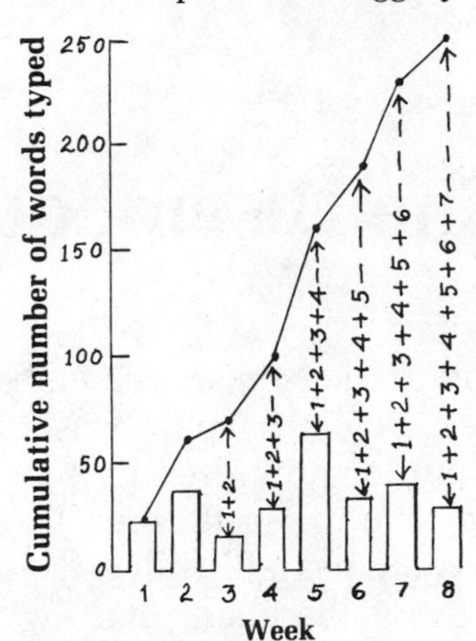

Fig. 4.13 Deriving a bar graph from a cumulative frequency chart.

TECHNICAL NOTES ON CHAPTER 4

On Reading Graphs: In a bar or line graph, there are two axes, the horizontal and the vertical. Commonly the horizontal axis is referred to as the *x*-axis; the vertical as the *y*-axis.

The *x*-axis is usually reserved for variables about which we are presenting quantitative information on the *y*-axis. Some typical variables shown on *x*-and *y*-axes are:

***x*-axis**	***y*-axis**
Identity of salespersons	Dollar volume of sales
Days of the week	Closing prices on stock or commodity exchanges
Dates (months, years)	Crime rates, population figures, industrial production, various indexes, dollar volume of sales
Categories of crime (burglaries, murders, rapes, etc.)	Number, proportion, or percentage committed in some time period.
Size of family	Frequency in some population of interest.
Post position on an oval race track	Number of wins
Test scores (IQ, aptitude, achievement)	Number of individuals

CHAPTER **5**

The Disembodied Statistic

Many of us old-timers remember the soap ad that proudly proclaimed, "Ivory Soap (or was it Ivory Snow?) is ninety-nine and forty-four one-hundredths percent pure." I must admit that I was most impressed when I first heard it. Now, no longer an impressionable and callow youth, I am more impressed by the audacity of the claim than by the purity of the soap. This ad beautifully illustrates what I call the *disembodied statistic*. The three essential ingredients of a disembodied statistic are as follows: (1) It must be exceedingly precise; (2) it should provide no frame of reference that allows the reader to make comparisons; and (3) the definition of the main concept should be left ambiguous.

Let us face it. Nothing blows the mind more and is as unassailable as a precise statistic. Ninety-nine and forty-four one-hundredths percent pure! That's hard to beat. But I must be honest with you. I have pulled a few classics in my time. On the average of 1.73562 times each year, a student will interrupt my lecture with a question such as: "Professor, is it true that we use only ten percent of our brain during the course of a lifetime?" Now I ask you, how do you answer a question like that? For years, my approach has been to break down that statement systematically into its component parts. What do you mean by *use*? How does a person use his or her brain? Unless we can come up with some satisfactory definition of this word, the rest of the statement is sheer gobbledygook. Even assuming that we can arrive at some mutually satisfactory definition, we're not much better off. How do we measure the proportion of the brain that gets "used" in a lifetime? Could we stick thousands of electrodes on various parts of the brain to determine whether the underlying nerve cells are firing? Ah, but it is a neurophysiological fact that they will all be firing if they are not dead, that is, 99.650374109 percent of the time. Hey, we could go on with this analysis *ad nauseam*. I think the point is clear. The

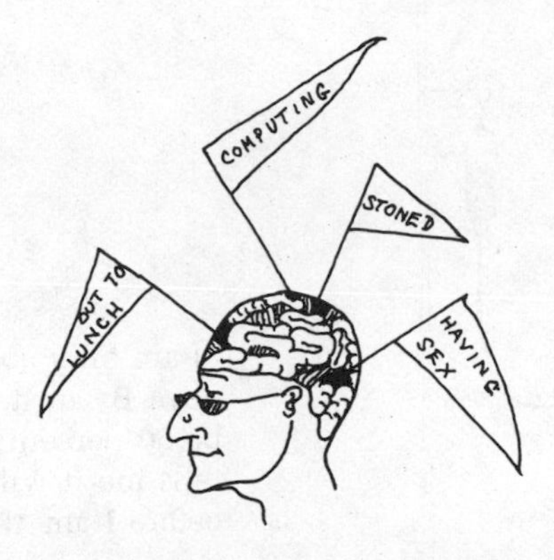

question is unanswerable and any attempt to provide a statistic is fraudulent, no matter how well intended.

How do I field this question nowadays? Why, I simply reply, "Your statement is absolutely false. The truth is that we use our brain only 8.45603 ± 0.000005 percent of the time." After giving the class a few moments to recover its composure, I continue: "I think that the question is really concerned with the efficiency with which we use our time and talents. The question of human efficiency is subject to some determination, within broad limits and under certain agreed-upon definitions of inefficiency." I then discuss various ways of measuring human performance.

A Few Examples

We are assailed by disembodied statistics with such frequency each and every day of our lives that I fear they have destroyed our critical faculties. We hear them, feel a bit uncomfortable about the meaning or accuracy of the claim, take a sip of beer (made with 99.44 percent pure water; the remaining 0.56 percent is a soup of additives and preservatives), and permit ourselves to be mesmerized by the dancing images on the boob tube.

- The star athlete, recovering from a disabling injury, proclaims with sickening sincerity, "I am only about forty percent right now. By next week, I expect to be sixty percent, but my doctor tells me it will be another month before I am a hundred percent." The TV commentator (not usually the most perceptive

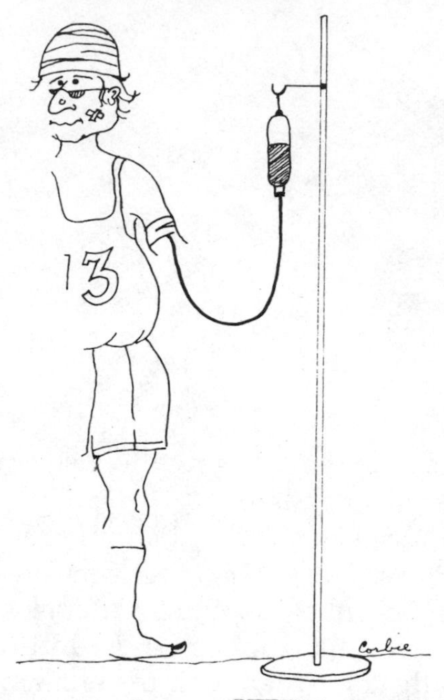

"I am only about 40 percent right now. By next week, I expect to be 60 percent, but my doctor tells me it will be another month before I am 100 percent."

person in the world) is delighted. "You heard what he said. He's only forty percent but a hundred percent of his opponents quake at the mere sight of him. Just wait until he's back to a hundred percent. Oh, boy!"

- The shill on "WeeWee Hours Movie Greats" proclaims "Dodoes are more effective." And I cry out from the depths of my insomnia, "Than what? What do you mean by effective?" But he never answers and my protests are lost in the ether.
- The Bureau of Labor Statistics is fond of making such pronouncements as "It costs fourteen thousand two hundred and fifty-one dollars for a family of four to enjoy a 'moderate' standard of living." No one can argue with the precision of the statement—it is precise down to the last dollar—but what is meant by moderate? Odds are that a thorough investigation would reveal that moderate is defined as "a standard of living that can be maintained on an income of $14,251."* Do I detect some circularity here?

* An investigation was conducted by the editors of *Fortune* magazine on just such a claim by BLS in 1967. Back when a moderate standard of living for a family of four could be bought for $9,194 they concluded that "moderate living standard" was defined as "a living standard that could have been bought by a family for $9,194." "Shadowy Statistics" (Editorial), *Fortune*, December 1967.

• I just went through a magazine and found advertising for eleven different brands of cigarettes. They all report the number of milligrams of tar and nicotine per "average" cigarette. But what does this tell you about the risk to your health? Moreover, assuming that lower quantities of tar and nicotine are less dangerous than higher quantities, the reader of the magazine would have to compare all the ads (as I have done in table 5.1) before deciding on his or her own brand of poison. Incidentally, I have heard a rumor that a new filter has been developed that completely blocks all tar and nicotine. It also blocks smoke. The only trouble is that 41.67 percent of the people trying to draw on the cigarettes become hernia victims.

The claims of advertisers, shills, and politicians remind me of a story I heard years ago in a Sunday sermon. It seems that there is some fabled restaurant district of some large American city (also fabled, I presume). One restaurant proclaimed in large letters, "THE BEST RESTAURANT IN THE CITY." The next on the block would not be outdone; "THE BEST IN THE STATE," it boasted. A third, larger and more garish than the other two, boldly stated, "WE ARE THE BEST RESTAURANT IN THE

Table 5.1 Amounts of tars and nicotine reported in separate ads in a single magazine. This list was compiled by me for your benefit. "Now that I have these statistics, what should I do with them?" you ask. I won't touch that question.

Brand		Mg/average cigarette Tar	Nicotine
Eve			
	Filter	18	1.3
	Menthol	18	1.3
Kent		18	1.2
Kool		17	1.2
Newport			
	Kings	18	1.2
	100's	20	1.5
Raleigh			
	Extra Mild	14	0.9
	Filter Kings	15	1.0
Saratoga			
	120's	16	1.1
	100's	16	1.1
Tarryton			
	Kings	20	1.3
	100's	19	1.3
True			
	100's Menthol	12	0.7
	King Regular	11	0.7
Viceroy		17	1.1
Virginia Slims			
	Regular	17	1.0
	Menthol	17	1.0
Winston		15	1.0

UNITED STATES." A fourth laid claim to the world's title. At the very fringe of the food-for-fun-or-ptomaine district stood a tiny, unimpressive hole in the wall. It displayed a diminutive sign in its window:

the best restaurant
on this block

But I think the real culprit in the growth of the disembodied statistic is not the advertising agency, the unscrupulous businessman, or the mass media. The fault is in us. All too often we demand precise statements and we will not settle for less. There is something comforting and reassuring about a number, no matter how questionable the assumptions from which it was derived. It's comforting to the person perpetrating the number (it makes you look so sagacious) as well as to the person who is happy to add your magic number to his or her arsenal of disembodied statistics (sometimes called trivia by the unappreciative).

Let me illustrate, again from my own life. Quite a few years ago, a close friend and chemist (Dr. Lawrence Rocks) and I wrote a book called *The Energy Crisis* (Crown Publishers, 1972). In it, we engaged in a number of broad statistical exercises with energy data and made some cautious (our words) extrapolations into the future. Early reviewers referred to these extrapolations as "alarmist." As one prediction after another was confirmed in the real world, later reviewers referred to these same projections as "overly conservative." We also made a number of somewhat less cautious political extrapolations based on our understanding of the facts of international politics. (For example, we stated that the Arab nations would use the conflict with Israel as a pretext for declaring an embargo on oil. The result would be exorbitant costs for petroleum products, inflation, and unemployment, particularly in the automotive industry.) At first, this book was welcomed and acclaimed with as much enthusiasm as a streaker at a Billy Graham revival.

But then things began to happen in the Middle East. Suddenly, but for a short period of time, Larry and I became genuine 62.53 percent celebrities. We received calls almost daily from TV and radio stations asking us to submit to interviews. Except for a few shows, the TV circus was unreal: "Dr. Runyon, tell us what the energy crisis is all about (you have one minute)," or "Dr. Rocks, what is the future of alternative energy sources? (Please keep it under thirty seconds.)" The radio shows were more relaxed, and permitted us to answer in considerable depth. In fact, I was on one show for five hours nonstop. But throughout all the interviewing, one demand persisted, either directly or by implication: "Give us precise statements! We like 99.44 percent. It sounds knowledgeable." Finally, on one radio show, I couldn't take it any more and gave a very precise statistic. The following is a general account of one segment of the interview. (It is not exact, since I do not have a tape recording of that show now. To tell you the truth, I can't even remember the name of the show.)

"Professor Runyon, tell us what percent of our total energy we can expect to derive from the sun by the year 1990."

"Well, that is a difficult question. The answer depends on so many factors—how much money the government allocates to solar research, the quality of research, the economics of providing solar energy—"

"Excuse us for interrupting, professor, but let's cut through the hedges. I would like to know, and I am sure our listeners would also like to know, just how much solar energy we can expect to have by 1990."

"Well, that's what I was getting to."

"I understand; but let's cut through all the ifs, ands, and buts. How much can solar energy contribute to our nation by 1990?"

"Under the right conditions, we could have 99.44 percent of our total energy from the sun." (I actually used the Ivory Soap figure; I really did!)

"Why, that's almost one hundred percent! That's the highest estimate I have heard yet. You really believe this is possible?"

"Absolutely, given the right conditions."

"Amazing. What are the right conditions?"

"We need a genuine biblical miracle."

To any among you who may have been one of the 27.463 people listening to that interview, I want to apologize here and now for my apparent flippancy.

The energy crisis—like so many crises that have surfaced over the past decade—has already built up an impressive repertoire of disembodied statistics.

Let's look at one example. While I was participating in an energy conference at the Watergate complex in 1973, one government official estimated that there are about 240 billion barrels of oil off the eastern continental shelf of the United States. That's a lot of oil. If recoverable, it would amount to an approximate forty-year supply at our present rate of consumption. Later, in conversations with oil company executives, we obtained estimates ranging from 5 to 40 billion barrels. The truth of the matter is that nobody knows. Most of the estimates are based on an analogy with other oil-producing sites sharing certain characteristics in common with the outer continental shelf. At worst, these estimates represent wild speculative guesses; at best, they also represent wild speculative guesses.

Buy why do these guesses differ by so much? If you were to accept the government official's estimate, you would have to agree with Ralph Nader (he has a statistic for *everything*) that we are drowning in oil. If you accept the least optimistic oil company estimates, you must wonder if it is worth the trouble and the financial risk to go after the gooey stuff. When you look a little deeper, you realize that there are two basic reasons for wide disparities in the estimate of "oil out there." One involves definition and the second is motivation.

What do we mean by "oil off the outer continental shelf?" One definition is "All the oil that is presumed to be out there somewhere." This is a very convenient definition for the government official to use because he is hoping to sell those leases to the oil companies (motivation). A figure of over 200 billion barrels sounds like something worth going after. But the oil executive rules out this definition. He's going to have to pay for the leases. It will break his consumer-oriented heart if he must pass on the increased cost to the person to whom he trusts his car. He says, "There's no way we will ever find all of the oil under the ocean even if it's there. Some will be too deep; some will be in small deposits that we'll never locate; some will not be economically recoverable, even if we find it. Why, even the best land-based wells rarely yield more than thirty percent of their contents." Quite naturally, the oil company executive plays the part of the reluctant bridegroom. He wants to get the best possible deal on oil leases. But even he must be careful to play the numbers game with consummate care. If he sets too low an estimate, the environmentalists will clobber him in the courts. They will show that the desecration of the seascape is too high a price to pay for so little oil. Table 5.2 shows how a whole lot of oil can be either a whole lot of oil or a mere dribble, depending on which definitions of "oil out there" you are willing to take.

No wonder the American public, the Congress of the United States, and the person who pumps gas at the corner station are confused by the energy crisis, the population explosion, the crime rate, additives in food, the environmental crises, resource shortages, and the thousand and one (notice my precision—it feels so good) issues that vie for our attention these days.

Table 5.2 How a government official, oil industry representatives, and an environmentalist can arrive at widely discrepant disembodied estimates of oil off the eastern continental shelf. Different definitions involve: total oil, oil that can be located, oil that can be recovered at reasonable (by whose reason?) economic cost.

Estimator	Estimates
Government Oil Official	240 billion barrels of oil off the Atlantic continental shelf. Enough to last our nation 40 years.
Optimistic Oil Official	We'll find half of it (120 billion barrels) and recover one third. That gives us 36 billion barrels, or enough to last our nation 6 years.
Pessimistic Oil Official	We'll find one-fourth of it (60 billion barrels) and recover 25% of what we discover. That gives us 15 billion barrels at the upper limit. But we think there are only 118.5 billion barrels to begin with. That means we can expect to recover only 7.40625 billion barrels. That may last us 1.1901 years or 14.28 months.
Environmentalist	They'll find one tenth of it, at great cost to the environment. Of the 24 billion found, they'll recover one fifth. That's less than 5 billion barrels of recoverable oil. Not even one year's supply.

Now, having shown that the disembodied statistic can apply to such unimportant things as energy and the environment, let us return to vital issues like clean and pure soap. What exactly have we learned when we are told that something is 99.44 percent pure?

First, I guess, we had better find out what "pure" means. In the dictionary I consulted, I found over twenty different definitions including: "A mixture—as of dogs' or pigeons' dung in water—for bating kips or skins after liming." I haven't the foggiest notion of what a kip and liming are, and I am too lazy to look them up. And bating sounds obscene. However, I doubt that pure was being used in this sense. But let us stipulate that we have arrived at a satisfactory definition of pure. What does 99.44 percent pure tell us? How does the purity of this soap compare with that of competing brands? For all we know, 99.44 percent may be a rather low level of purity. Maybe a Brillo pad is 99.86 percent pure. Should I then wash my face with Brillo? And while we are on it, what does purity have to do with the

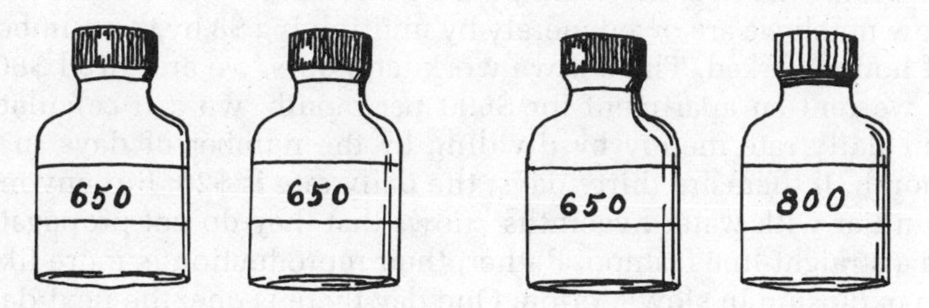

cleansing powers of soap? Maybe a soap that is 52.53 percent pure does a better job of cleaning. I don't know, and certainly the ad doesn't help me decide.

Nor was I told much when Volvo informed me that 90 percent of their cars sold in America over the preceding ten years are still on the road. (However, I suspect that very few Volvos were sold here ten years previously. My guess is that most U.S. Volvos were of rather recent vintage. It would be a supreme disgrace if nearly 100 percent of these were not still in service.) Nor when I see that graph showing that Anacin (or is it Bufferin or Excedrin? I get them all mixed up), with a higher level of pain reliever, gets into the bloodstream faster and maintains a higher level of pain reliever longer than competing brands. Why not just take more of the chain-store brand of aspirin and enjoy the same high level of relief at a lower cost?

Do you want to bring fun back into TV viewing? I recommend you forget about the programmed stuff and concentrate your attention on the commercials. But don't be offended when you suddenly realize that someone out there thinks you have the intelligence of a three-year-old chimp.

Extrapolating Exponential Growth

I am sure you have heard this puzzler. A single water hyacinth is placed in a pond. Each day the number doubles so that at the end of twenty days the entire surface of the pond is covered. How long do the hyacinths take to occupy one half of the pond's surface?

Our first tendency is to blurt out "ten days." This is due to the fact much of the arithmetic of daily living involves straight-line relationships. If we are paid $8 an hour, we can determine how much we are owed merely by multiplying $8 by the number of hours worked. Thus, if we work ten hours, we are owed $80. If we rent an apartment for $600 per month, we can calculate our daily rate merely by dividing by the number of days in a month. If there are thirty days, the daily rate is $20. But anyone familiar with water hyacinths knows that they do not propagate in a straight-line fashion. Rather, their reproduction is more like an explosion in slow motion. One day there is one; the next day there's two; the next day, four; the succeeding day, eight; and so forth. Why, last summer I introduced two into a fish pond with approximately 200 square feet of surface. Within less than a month, the entire surface was cluttered with hyacinths.*

The answer to the puzzle? Why, nineteen days, of course. If the number of plants doubles each day, then on the twentieth day the number is double that of the nineteenth day. Consequently, compared to day twenty there are half as many hyacinths in the pond on day nineteen. In fact, there were only one-quarter as many on day eighteen, one-eighth as many on day seventeen, and so forth.

Here's another one you have surely heard. A down-and-out baseball player went to a big-league manager at the beginning of spring training. Joe Aintgotit explained his burning desire to make the big time and offered a contract his manager couldn't refuse. "I'll play for a penny the first day, two pennies the second day. I only ask that you double my salary each day for thirty days. In other words, I want only a 30-day no-cut clause in which I am willing to give my all for just pennies a day. In fact, you may dispense with the usual room-and-board allowance. That's how much an opportunity to make the team means to me. At the end of thirty days, we can negotiate the contract, if you feel I've got what it takes."

The manager happily signed the contract. He was ecstatic at the unprecedented opportunity to get a "free" look at a highly motivated athlete. Only, much to his chagrin, he discovered that his perfect "deal" involved a somewhat greater financial risk than he had bargained for. To be more precise, he paid out

* Here's a little-known fact about hyacinths. They are fantastic garbage collectors. They remove many pollutants from the water. Look for garbage treatment plants in the future to include hyacinth ponds as garbage digesters, particularly in warm climes.

$5,368,709.12 on the thirtieth day; $2,684,354.56 on the twenty-ninth day; $1,342,177.28 on the twenty-eighth day; etc. Naturally, the manager was fired by his new owner—Joe Aintgotit, who, sad to relate, never made the club.

That's the main characteristic of types of growth that we call exponential. One day they are hardly discernible; the next day, month, year, they explode in your face. Some everyday phenomena that pursue roughly exponential growth rates are population, compound interest, consumption of resources, and development of new industries (e.g., petroleum, auto, plastics, computers).

The fun comes when we attempt to extrapolate these growth rates to make precise projections about the future. There's a little trick in making these projections that everyone should master. If you know the rate of compound interest, you divide that into 70 and obtain a rough approximation of the doubling time. Let's look at a few examples.

- At the age of 30, you deposit $1,000 in a savings bond that provides 10 percent annual rate of interest. Assuming that the bond is indefinitely renewable at the same rate of interest, how old would you be before you doubled your savings? Well, seventy divided by 10 is 7. Seven years is the doubling time. Thus, by age 37, you would be the proud possessor of a bond worth $2,000. By 44, it would be worth $4,000; $8,000 by the age of 51. "What's that question? How much would be in that account by the time you reached the century mark in age? That's easy. Over a million bucks. Why that dour look? Don't think you'll make it to a hundred? Oh, it's not that at all? Oh, I see, you're afraid inflation will take the edge off being a millionaire." (See box 5.1.)
- The present world population is about 4.376 billion. The compound rate of growth is roughly 2 percent per year. That means the doubling time is around thirty-five years. Does that mean the world's population will be 8.758 billion in 2015, 17.516 billion in 2050, and about 35 billion by 2085? It could, if it were possible to maintain the 2 percent rate of population growth! Is this a reasonable expectation? Are we likely to be able to quadruple food production, resource consumption, and

Box 5.1

"Say, Dick."

"Yes, Hal?"

"I hate to smother your friend's joy. That rule of thumb of dividing the rate of compound interest into seventy is OK as a crude approximation."

"Of course. That's what I said."

"But when you start compounding a slight rounding error over many doubling times, the disparity can become quite large."

"Like for instance?"

"That poor guy wouldn't really have a million bucks by his centennial year. He would have only seven hundred eighty-nine thousand seven hundred and forty-six dollars and ninety-six cents."

"Really? How do you figure that?"

"I have a pocket calculator that has a button labeled y^x. How about you?"

"Why, yes. I wondered what that was for. How do you use it?"

"Put in the rate of interest, preceded by a one."

"Wait a minute. Pass that by me again."

"A five percent rate of interest would be entered as 1.05; ten percent as 1.10; fifteen percent as 1.15; and so forth."

"I get it."

"Let's do ten percent. So you enter 1.10."

"Got it."

"Now, hit the button y^x."

"Right."

"Now enter the number of years you want to compound the ten percent interest."

"Let's do seventy. That's when our friend will be a hundred years of age. OK?"

"Hit the *equals* button. What do you get?"

"I get 789.7469568. I see what you mean."

"Right. When you multiply by a thousand you find the amount of money he will have in the savings account when he reaches the century mark."

"Good. What do you do if you don't have a calculator?"

"I thought you'd never ask. I have prepared a table [table 5.3] that shows the doubling time for compound rates of interest from one percent through twenty percent. It also shows the amount of money a dollar would accumulate after various numbers of years at the rates of interest shown."

"Thank you, Hal. I'm sure our readers will find this very interesting."

"And useful. It also shows halving time."

"I think you had better explain."

"Well, let's assume that the annual rate of inflation is twelve percent. If continued at that rate, a dollar would purchase only half as much goods and services after 6.1 years. If your present income is twenty thousand, you'll require forty thousand in about six years just to keep pace with the inflationary rate."

"That's a sobering thought."

all of those other goodies in little more than a century? And what about the year 2120? Sixteen times as many people as today? I'll bet anyone one million bucks that this will not happen. Any takers?

• For a decade prior to 1970, the energy growth rate in the United States was about 7 percent per year. Based on that rate, the doubling time in energy use would be about ten years; the quadrupling time would be about twenty years. In *The Energy Crisis*, Dr. Rocks and I suggested that by the turn of the century we could be using *eight* times as much energy as in 1970. Inveterate hedgers that we are, we followed up this startling projection with the disclaimer, "Obviously, this compounded rate can't be maintained indefinitely."[1] Indeed, it has not. Thanks largely to the Organization of Petroleum Exporting Countries and declining oil reserves in this country, the rate of increase in energy use has dropped sharply. It is now less than 2 percent per year, compounded. At the present rate, we would estimate the quadrupling time as seventy or more years in the future. How fragile are predictions from unstable growth rates!

Of course, they must be. The ready availability and accessibility of an important resource invites its exploitation as rapidly as possible. But then less becomes available. What is available becomes less accessible, and the compounded growth rate is sharply curtailed. The $64,000 question then becomes, "Will we find viable alternatives to head off the inevitable social, political, financial, and personal upheaval that must follow a failure to do so?" Although I cannot prove it, I have a hunch that many advanced civilizations that have mysteriously disappeared in the past have done so because they exhausted some key elements for which they could find no alternatives. Thought for the day: Will energy be our Achilles' heel?

[1] Rocks, L. and R. P. Runyon, *The Energy Crisis*, 1972 (Crown, p. 10).

Table 5.3 Component Interest and Doubling Time

Rate of Compound Interest (percent)	Doubling Time (years)	A Dollar Compounded After Varying Numbers of Years at the Rate of Interest Shown							
		1	3	5	7	9	11	13	15
1	69.7	1.01	1.03	1.05	1.07	1.09	1.12	1.14	1.16
2	35.0	1.02	1.06	1.10	1.15	1.20	1.24	1.29	1.35
3	23.4	1.03	1.09	1.16	1.23	1.30	1.38	1.47	1.56
4	17.7	1.04	1.12	1.22	1.32	1.42	1.54	1.67	1.80
5	14.2	1.05	1.16	1.28	1.41	1.55	1.71	1.89	2.08
6	11.9	1.06	1.19	1.34	1.50	1.69	1.90	2.13	2.40
7	10.2	1.07	1.23	1.40	1.61	1.84	2.10	2.41	2.76
8	9.0	1.08	1.26	1.47	1.71	2.00	2.33	2.72	3.17
9	.1	1.09	1.30	1.54	1.83	2.17	2.58	3.07	3.64
10	7.3	1.10	1.33	1.61	1.95	2.36	2.85	3.45	4.18
11	6.6	1.11	1.37	1.69	2.08	2.56	3.15	3.88	4.78
12	6.1	1.12	1.40	1.76	2.21	2.77	3.48	4.36	5.47

13	5.7	1.13	1.44	1.84	2.35	3.00	3.84	4.90	6.25
14	5.3	1.14	1.48	1.93	2.50	3.25	4.23	5.49	7.14
15	5.0	1.15	1.52	2.01	2.66	3.52	4.65	6.15	8.14
16	4.7	1.16	1.56	2.10	2.83	3.80	5.12	6.89	9.27
17	4.4	1.17	1.60	2.19	3.00	4.11	5.62	7.70	10.54
18	4.2	1.18	1.64	2.29	3.19	4.44	6.18	8.60	11.97
19	4.0	1.19	1.69	2.39	3.38	4.79	6.78	9.60	13.59
20	3.8	1.20	1.73	2.49	3.58	5.16	7.43	10.70	15.41

CHAPTER **6**

Ratios, Proportions, Percentages, and Peptic Ulcers

Imagine you are an advertising executive in charge of a very lucrative account. The sales line in the graph your boss shows you daily has taken on the appearance of a ninety-meter ski jump. You have tried everything to reverse the trend—worry, Di Gel, Bromo-Seltzer, and losing yourself in mindless TV viewing. Nothing helps.

Your mind fondles the good old days when you could fib the pajamas off a zebra and no one would be offended, telling yourself, "It's part of the game, like delivering a karate chop to the neck of the opposing quarterback when the officials are not looking." But, alas, those scoundrels in Washington want truth in advertising. (Except, naturally, for advertising done by the federal government.) During World War II we were assaulted with spot commercials on radio commanding us to buy war bonds to "nip the Nips." Actually, the purchase of these bonds did not put tanks in the fields or planes in the sky. They merely took a lot of excess money out of circulation and thereby lowered the rate of inflation. (As of this writing, there is much talk about selling "energy bonds." Not a cent will go into energy, but widespread acceptance could lower the rate of double digit inflation.) Now, with truth in advertising, the radio, the tube, and the private sector in general can no longer engage in good clean honest-to-God lies. They are compelled to resort to guile, deception, and numerical legerdemain. Oh, how the government makes cheaters of us all!

The Weasel Third Category

So how do you deceive without telling an out-and-out lie? Try calling on your head of marketing research.

"Dolores, we're in trouble. Brand Kissyours is grabbing the lion's share of the market. We're relying on your department to get us out of the morass."

"No sooner said than done, boss. We have just completed a study that's sure to put Kissmy on top. The report's going to Advertising this afternoon. From then on, it's their baby to run with."

"You got good stuff?"

"Yes."

"Why am I the last to hear about it?"

"We were going to present it this afternoon."

"Well, let's have it right now." You reach into your pocket for a Di Gel and surreptitiously place it in your mouth while pretending to pick your nose. "What sort of comparison did you make?"

"A hugability test."

"Come again?"

"We had the subjects fondle, caress, and hug our brand of toilet tissue (Kissmy) and their brand. We asked them to judge which felt softer."

"Great idea. And we won?"

"Here's the data. We'll let you be the judge."

Dolores spreads the table of data in front of you. A pained look passes briefly across your face as you reach for another Di Gel. "You call this good? This shows only twelve percent of the subjects preferring Kissmy." Your lips curl back in the rictus of rage.

	Kissmy Preferred	Both About the Same	Kissyours Preferred	Total Sample
Number Selecting	24	96	80	200
Percentage Selecting	12	48	40	100

"Calm down, boss. These are just the raw data. Nobody outside the organization will ever see them."

"So what are you going to do? Fabricate results? What happens when the Feds ask to see our data."

"We'll show them the same chart that our TV viewers see."

"Yeah? By what magic are you going to make this sow's ear into a silk purse?"

"By strictly legitimate means, I assure you. We'll not lie, deceive, cheat, or fabricate. We'll tell it exactly like it is."

"I'm waiting."

"You see, we included the *Weasel third category*. Some call it the *swing category*. What it's called is of little importance. It's what it allows us to do legitimately that makes it so beautiful."

"I'm still waiting."

"Let me first give you a little background. We asked each person to tenderly caress each brand of bathroom tissue. They were packaged identically, so they could not tell which brand was which. They were then given three alternatives from which to choose: A (Kissmy) is better; B (Kissyours) is better; and C, Both are about the same."

"So where does this leave us? We have forty-eight percent undecided but still only twelve percent choosing Kissmy."

"So we combine categories."

"Come again."

"We combine categories. Forty-eight percent found both brands about the same and twelve percent preferred Kissmy. Adding them together, we can legitimately claim that sixty percent of those tested found Kissmy at least as soft and caressable as Kissyours." (See figure 6.1.)

"Brilliant! Absolutely brilliant!"

"Thank you. I didn't do it alone."

"And you're sure the Feds won't kick up a ruckus?"

"How can they? We're telling the absolute truth."

"I like it! Yes, I like it! What computer did you use to analyze the data?"

"We didn't use a computer. We did it all by hand."

"By hand?"

"Yes, with all this computer espionage going on these days,

Figure 6.1.

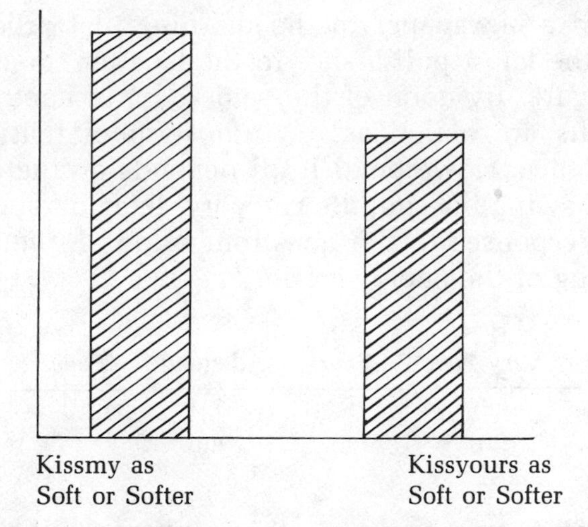

we couldn't take a chance that our data would fall into Kissyours's hands."

"Good thinking. That could be a problem."

"Absolutely. Why, using exactly the same data, they could show this chart" (see figure 6.2).

"Oy! Are you sure you destroyed all the raw data?"

"Sure. Oh, yes, by the way . . . about that raise you've been promising me . . ."

The use of the Weasel third category is not the exclusive domain of marketing and advertising agencies. Even the fourth estate has been known to fall victim to its allures. In my travels,

Figure 6.2.

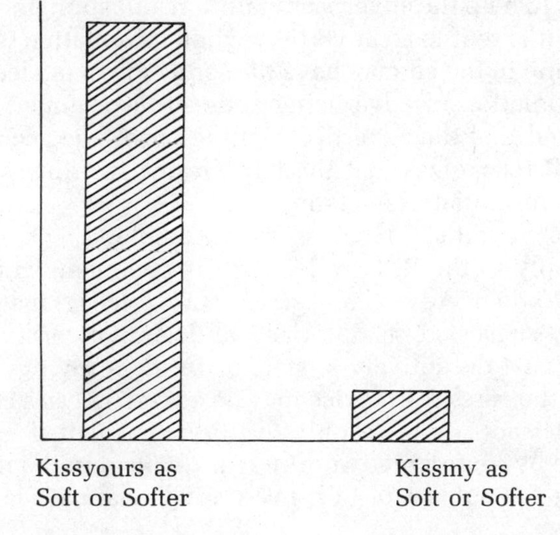

I have read a newspaper in the morning that proclaimed the results of the latest poll a slap in the face for some prominent political figure. By noon of the same day, in another city, the same results are hailed as a vindication of that politician's record in office. How come? It all depends on the direction in which you swing that middle category. Imagine you obtain the following responses to the question: "How do you feel about A's handling of the energy crisis?"

	Very Poor	Poor	Adequate	Good	Very Good
Percentage Responding	10	30	30	15	15

If you like A, you could say that 60 percent of those polled found his or her record satisfactory or better. If you are on the other side of the political see-saw, you could take consolation in the fact that only 30 percent found A's performance good or better. By implication, 70 percent were dissatisfied.

So, you see, percentages provide a golden opportunity to engage in statistical mischief. But you ain't seen nothing yet.

Box 6.1 A Word to the Wise

The Arizona Daily Star, March 25, 1980. Advertising Double Check. A BBB Service to Advertisers. The purpose of the Advertising Double Check is to help the advertiser maintain and strengthen customer confidence. It is sent to an advertiser when information is requested, when wording in the ad may have the tendency to mislead, or when there is a violation of advertising code or government regulation. It is suggested that the advertiser double check the accuracy of the ad. The BBB recognizes that the burden of proof lies with the advertiser to substantiate his claim.

When the advertiser receives the "Double Check" he should promptly reply to the BBB, reflecting his comments and/or action.

The BBB Code of Advertising states "List price," "manufacturer's list price," "suggested retail price" and similar terms have been used in the past deceptively to state or imply a savings which was not, in fact, the case. A list price may be advertised as a comparative to the advertised sale price only to the extent that it is the actual selling price by representative principal retailers in the market area.

Regarding the subject of "up to" savings claims, the BBB Code of Advertising states "Savings or price reduction claims covering a group of items with a range of savings shall state both the minimum and maximum saving. Savings claims involving items comprising less than 10% of the total items offered in the sale should not be included in the price range."

Despite an advertiser's best efforts to ascertain competitive prices, the rapidity with which prices fluctuate and the difficulty of determining prices of all sellers at all times precludes an absolute knowledge of the truth of claims such as "Lowest prices."

The Better Business Bureau has mounted a campaign to protect the consumer against misleading and fraudulent advertising claims. Have you ever gone into a store advertising up to 80 percent off and asked to see the merchandise with the huge markdown?

"Oh, we sold them all. They went first thing," says the salesman, averting his eyes.

"Really? What were they?"

"What were they?"

"Yes, what were they?"

"You know, I'm not really sure."

"If you're not sure, how do you know they have been sold?"

"Well, you don't see them, do you?"

"No, I don't. That's why I came to you."

"Why don't you talk to our sales manager? She's the lady standing over there in Underwear." Quick exit, accompanied by unusual solicitude for another customer.

"Ms. Sales Manager in Underwear, may I speak to you a moment?"

A detached distant look. No reply.

"Excuse me, but I have a question."

"Yes?"

"About those items on sale for eighty percent off—"

"All sold out."

"Already? The sale just started today and I was one of the first customers in the store."

"Oh, yes, they were special close-out items. We never received them."

"What were they?"

"Gismos."

"Gismos?"

"Yes, gismos." A haughty look on her face. "You know what a gismo is, don't you? Everybody knows what a gismo is."

"Well, of course I know. What do you think I am? And I want one!"

"I'll be glad to give you a rain check."

"Good. When do you expect them in?"
"Expect *what* in?"
"The gismos."
"Maybe tomorrow; maybe never."
"Never?"
"I told you. They were close-out items." The reappearance of the distant look. Suddenly an accelerated movement away from you to answer a long distance call from her boss in the Gulag Archipelago.

Percentages, Percentages Everywhere

Often data are presented in tabular form that cry out to be expressed either as a percentage or a proportion. Take a look at the following table. It shows the annual incidence of alcohol-related collisions between 1968 and 1973 (as of July 1 each year) among motorists in three age groups. The data were collected in London, Ontario, in our good neighbor to the north.* The authors' intention was to shed light on changes in alcohol-related accidents following the lowering of the legal age from 21 to 18 for the purchase and consumption of alcoholic beverages. The new law took place at the beginning of 1971–1972.

Now here's the problem. There are at least three different proportions or percentages you can calculate from this table. For starts, you can divide each entry by the total (17/606, 26/606, 27/606, etc.) and obtain a percentage for each entry. Figure 6.3 shows what its graph would look like.

		Age Group		
Year	**16–17**	**18–20**	**24**	**Total**
1968–1969	17	47	5	69
1969–1970	26	39	7	72
1970–1971	25	48	9	82
1971–1972	33	133	14	180
1972–1973	23	153	27	203
total	124	420	62	606

*Whitehead, P. C., J. Craig, N. Longford, C. MacArthur, B. Stanton, and R. Ferience, *Collision Behavior of Young Drivers* (*J. Studies of Alcohol*, *36*:1208–123, 1975).

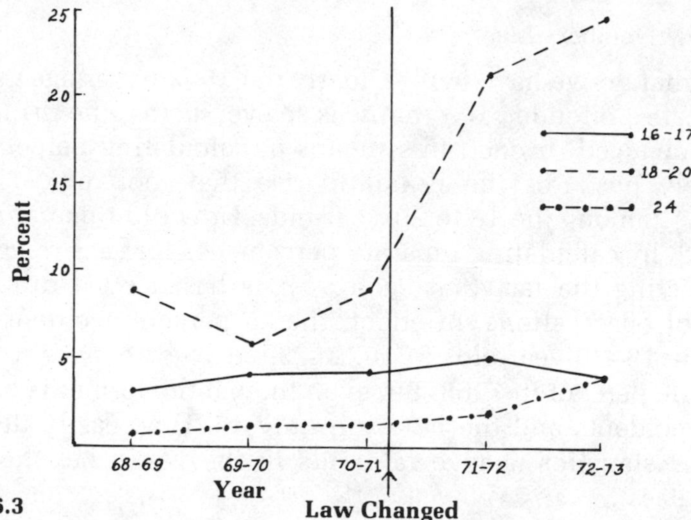

Fig. 6.3

'Tis clear, 'tis not? The alcohol-related collisions among those age groups not affected by the change (16–17 and 24-year-olds) showed a remarkable stable proportion of collisions throughout the five-year period. But the percentage among the newly liberated 18–20-year-olds goes into orbit.

Ah, but that's only percentage calculation one.

Mr. Obseuranti comes along and objects. "Wait a minute," he says, "I think we should find the percentage for each year and plot that."

"Why?"

"It truly reflects the percentage of change from year to year."

Although I disagree with the defense, we should present his case in the interests of fairness. Figure 6.4 shows what we find.

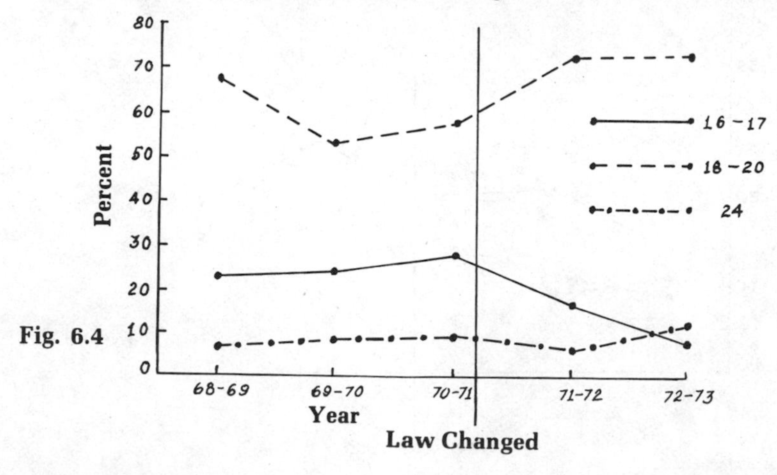

Fig. 6.4

Now what do we have here? Clearly the 18–20-year age group is the greatest offender. But that was so even before the drinking age was changed. In fact, this means of calculating percentage completely obscures the dramatic rise that took place after 1970–1971 among the 18 to 20-year-olds. How did this happen? In a word, by calculating separate percentages for each year, we are neglecting the fact that each year is based on a different number of observations. In effect, the 75 percent accident rate among the 18–20-year-olds in 1972–73 receives no more weight than the 68 percent in 1968–69, even though the former is based on 153 accidents and the latter on only 47. How easily do the perfidious statistics change raiments in the world beyond the looking glass!!

But let us not forget percentage three. This involves calculating a separate overall percentage for each age group by dividing each column total into each of the five entries that comprise it. To illustrate, the percentage of 16–17-year-olds who were involved in alcohol-related collisions in 1968–69 was:

$$\frac{17}{124} \times 100 = 14\%.$$

When we do this for each entry and plot in a graph, we obtain the results shown in figure 6.5.

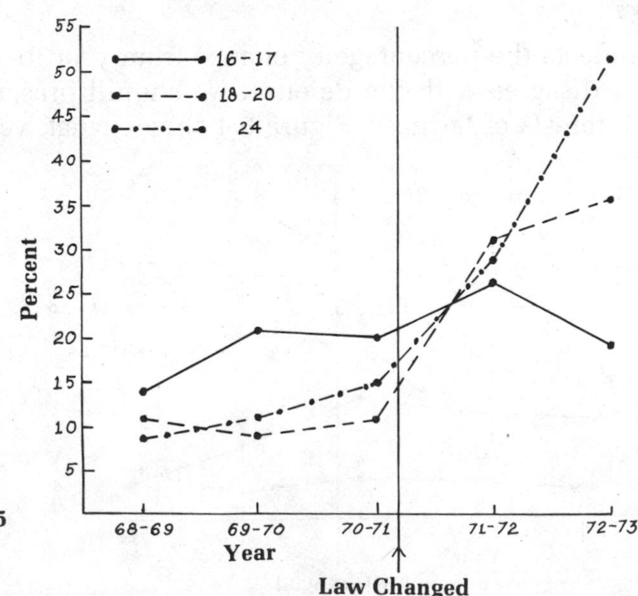

Fig. 6.5

Mirable Dictum! The 18–20-year-old group is suddenly lost in the shuffle as the 24-year-olds emerge as the genuine honest-to-God culprits. How can the same data give rise to such contradictory impressions? Is it true that the statistical world behind the looking glass is as chimerical, capricious, and arbitrary as some critics would have us believe? No. There's a reason for everything. By calculating separate percentages based on each column total, we are putting all three on the same scale (one hundred percent) and ignoring the fact that the total number of alcohol-related accidents in which the 24-year-olds were involved is only about one-seventh as many accidents as the 18–20-year-old-group.

Now, how do you go about deciding which of the three calculations represent truth in this case? Both percentage two and percentage three are seriously flawed, whereas percentage one is not. The choice seems clear unless you have a vested interest in proving one thing or another. All of which reminds us that statistics don't lie but people do.

Stealing by the Base

Baseball is not the only American pastime that involves both pilferage and a base. In fact, if you compare the exploits of Maury Wills with the accomplishments of hundreds of merchants spreading from coast to coast, the former Dodger infielder will appear like a bush-leaguer. You have all seen the ads: "BELOW COST SALE." And you wonder how they can do it and remain in business. Maybe they take a small loss on each item and make it up in volume? A great scenario for bankruptcy.

No, the whole flimflam hinges on the definition of "cost." Let's take a look at the manufacturer of flimsies and see how many ways "cost" can be defined.

Cost 1: The actual cost to manufacture the item. This includes cost of raw materials, labor, machinery, and overhead. Let's say cost 1 is $2.00 per item.

Cost 2: A distributor purchases flimsies from the manufacturers for $2.50 per item.

Cost 3: The retailer buys them for $3.50 each.

Cost 4: The retailer calculates shipping, taxes, overhead, advertising, and costs of sales and finds that the real cost is $4.00.

Cost 5: The retailer calculates a reasonable profit—50 percent—and marks the price up to $6.00.

Cost 6: However, the retailer is planning a big storewide sale in two weeks. The price tag is labeled $8.00 to provide some room for downward maneuvering. (This is absolutely critical! To advertise a sale, you must show a price decrease.)

Are you beginning to get the picture? With a half-dozen different costs to choose from, the advertising menu is truly a smorgasbord. How much below cost do you want to advertise? You can take your pick.

The general formula for calculating the percentage of markup or markdown is

$$\text{Percentage markup or markdown} = \frac{\text{Cost B} - \text{Cost A}}{\text{Cost A}} \times 100$$

The bottom of the equation (cost A) is known as the base. It is the selection of this base that makes possible a bunch of statistical shenanigans. Let's look at some of them.

Let us say you sell a product for $8 (cost B) for which you paid $4 (cost A). Your percentage of markup is

$$\text{Percentage markup} = \frac{\$8 - \$4}{\$4} \times 100 = 100\%$$

Not bad. What's that you're saying? Very bad? How come?

"I couldn't sell enough at that price to make the slightest dent in inventory."

"So, what do you do?"

"I reduce the price by two dollars and advertise a below-cost sale."

"Oh, how does that work?"

"Well, my usual cost is eight dollars, right?"

"Right."

"And my new cost is six dollars."

"So, I'm selling for two dollars below my usual cost. In fact, using the formula and using eight dollars as a base, the percentage reduction is

$$\frac{\$6 - \$8}{\$8} \times 100 = -25\%$$

"That's a twenty-five percent loss. I'm going to the poorhouse just to please my customers."

"What do you mean you're taking a loss? Even at the new price, your markup is fifty percent."

"Yeah. And what was my markup before?"

"A hundred percent."

"And now, it's fifty percent. Right?"

"Right. So what are you complaining about?"

"Using fifty percent as my base, I am taking a one hundred percent loss. Figure it out for yourself. My profit is now fifty percent. We'll call that B in the formula. My profit before was one hundred percent. We'll call that A. So, using the formula

$$\frac{50\% - 100\%}{50\%} \times 100 = -100\%$$

"You see? I'm giving the merchandise away. How generous can a guy get?"

"I have the uneasy feeling I am being put upon by a premier flimflam artist."

"I am deeply offended. What I am doing is all aboveboard and legal."

"Now, let me get this straight. When you were charging eight dollars, your markup was one hundred percent?"

"Right."

Box 6.2

"Enough! Enough!"

"What's the trouble, Hal? A bit of indigestion?"

"You hit it right on the head."

"Something you ate?"

"Right again. I'm trying to digest this stuff about bases. If you're succeeding in confusing me, I'm sure I have a lot of your readers for company."

"Well, I explained it as well as I could."

"Not good enough. What you need is a visual display."

"Like a table?"

"Yes. I have prepared a little table that summarizes it all. All you need do is look up the retailer cost in the row and the sales price in the column. At the conjunction of the two is the percentage markup or loss. Pretty nifty, eh?" (See table 6.1)

Table 6.1 Percentage of markup or loss at varying costs to retailer and varying sales prices. The rows represent the retailer's actual cost and the columns, the sales price. The entries in the cells represent the percent markup or loss rounded to the nearest percent. The formula used is [(Sales price − cost) ÷ cost] × 100. Thus, if a retail establishment pays $2 for an item (row 1) and sells it for $8 (column F), its markup is 300%, or three times cost. If it pays $4 for an item (row 4) and sells it for $2 (column A), it shows a loss of 50% and is headed for Chapter 11 bankruptcy. But what if it pays $2 for an item, sells it for $4, but claims $8 as its cost basis? Its actual markup is 100% but it *claims* it is selling at 50% below cost. This is truly stealing by the base.

	Sales Price					
Actual Cost to Retailer	**A $2.00**	**B $2.50**	**C $3.50**	**D $4.00**	**E $6.00**	**F $8.00**
Cost 1 $2.00	0%	25%	75%	100%	200%	300%
Cost 2 $2.50	− 20%	0%	40%	60%	140%	220%
Cost 3 $3.50	− 43%	− 29%	0%	14%	71%	129%
Cost 4 $4.00	− 50%	− 38%	− 12%	0%	50%	100%
Cost 5 $6.00	− 67%	− 58%	− 42%	− 33%	0%	33%
Cost 6 $8.00	− 75%	− 69%	− 56%	− 50%	− 25%	0%

"When you reduced the price by two dollars, you claimed that you were selling at twenty-five percent below cost."

"It's not a claim. It's the exact truth."

"But your profit was still fifty percent."

"True. But that means I was taking a one hundred percent loss."

"In other words, a profit of fifty percent is, at once, twenty-five percent below cost and a profit loss of one hundred percent?"

"Well, Hank, that's exactly right. Now you got it. Everything depends on the base. Now let me tell you how I can take a two hundred percent loss and still turn a profit . . ."

Paradoxical Percentages

My very favorite of all misleading statistics is what I call *paradoxical percentages*. When I first discovered* them a number of years ago, and excitedly explained them to a colleague, he exclaimed, "Why, that's impossible!" I'll let you be the judge.

Let's put your imagination to the supreme test and suppose that you have been hired as a statistical consultant to a school district. The superintendent explains the problem: "We have been accused of sex discrimination in our hiring practices. On the surface, the evidence appears quite convincing. Over the past ten years, we have received applications from six hundred male and six hundred female teachers. We hired a hundred and thirty-three of the males and a hundred of the females. In other words, contracts went out to 22.2 percent of the male applicants [$133/600 \times 100 = 22.2\%$] and only 16.7 percent of the females [$100/600 \times 100 = 16.7\%$]. One of our administrators recently completed a graduate statistics course, and she says that the difference is statistically significant, whatever that means."

"The term statistically significant means that the differences are not likely to have resulted from chance. They are more likely

* I am not laying claim to being the original discoverer. I have an annoying habit of making a remarkable discovery, only to find later that someone had already done it 200 or 300 years before. I'm not much more successful as a gold prospector. I am forever finding "promising sites" that I later find are already staked or rich in fool's gold.

to have arisen from nonchance factors, such as discriminatory hiring practices."

"Come again?"

"Let's say you flip a coin one hundred times and come up with fifty-three heads and forty-seven tails. Would you be willing to accuse the coin of being biased in favor of heads?"

"No."

"Why not?"

"It's so close to fifty-fifty, which I would expect by chance."

"Good. But what if it comes up ninety heads and ten tails?"

"I'd want to take a pretty close look at the coin or the way you are throwing it."

"Why?"

"I don't know much about probabilities, but I'm willing to bet that there is very little likelihood that it occurred by chance."

"Precisely. The first outcome, fifty-three to forty-seven, is not statistically significant. We're willing to assume the coin is honest. In the second case, ninety to ten, the difference is statistically significant. In fact, the likelihood that the result occurred by chance is considerably less than one in a billion. Now we're not so keen about the idea of saying the coin is honest. We must look elsewhere for an explanation."

"I see. Since the difference in the percentage of males and females hired in our school district is statistically significant, we can pretty well rule out chance as an explanation. Based on historical precedent, a reasonable explanation for the disparity is discriminatory hiring practices. That being the case, is there further need for your services?"

"Of course," you quickly reply, after extracting your foot from the region of your orbicular oris muscles. "When we dig deeper into the data, we may uncover some reasonable explanation for the apparent favoritism of males over females." You had better say something like that. Statistical consultantships are not easy to come by. Those who talk themselves out of jobs rapidly become as extinct as the dodo.

Having been hired, you are suddenly thrust into a roiling sea of data. Twelve hundred personnel folders are unceremoniously dumped into your lap. What is a person to do? Organize them. Your first thought is to group them into two piles—males and females. Then you subdivide each of these into three categories:

grades K–8 (grammar school), grades 9–12 (high school), and grades 13–14 (two years at the community college). When you tabulate the number of applicants in each of these three categories, by gender, and the number hired, you obtain the following table:

	Male		**Female**	
Job Category	**Number of Applicants**	**Number Hired**	**Number of Applicants**	**Number Hired**
Grades 13–14	200	40	30	12
Grades 9–12	300	90	100	60
Grades K–8	100	3	470	28
Total	600	133	600	100

The first thing you notice is that the totals are in agreement with the superintendent's figures. You recalculate the percentages and find that these are also as previously presented. Then you calculate the percentages of males and females hired within each category. The following table results:

	Percentage Hired	
Job Category	**Males**	**Females**
Grades 13–14	20%	40%
Grades 9–12	30%	60%
Grades K–8	3%	6%

When you look at the figures, you do a double take. How can this be? The percentage of females hired within each category is *double* the percentage of males hired *in every category*. "It's impossible," you cry out in dismay. "I must have made a mistake," you whisper, in the event someone is eavesdropping. But quick calculations show you have made no errors. Thus, from precisely the same data you have found (1) a significantly greater percentage of males than females hired in the school district over the past ten years; (2) twice as great a percentage of females hired in each job category over the same time period. 'Tis a paradox, 'tis not?

Now, let's try to resolve the paradox. The basic problem with these data is that an equal percentage of males and females did not apply for each job category. A much greater percentage of males than females applied for grades 9–12 and 13–14, as you can see below:

Percentage of Males and Females Applying

Job Category	Males	Females
Grades 13–14	33%	5%
Grades 9–12	50%	17%
Grades K–8	17%	78%

Aha. Now it becomes clear. The majority of males (83 percent) were applying for jobs that were plentiful. Few females applied for the same jobs (22 percent). Even though the *percentage* of females hired in each of these categories was double the percentage of males, there were so few females applying that the total number of males hired was much greater. Not so the K–8 category. Fully 78 percent of the females applied for the *least* plentiful jobs. Thus, although their percentage edge over the males was still double, very few of either sex were hired. There's a moral here: Better to apply for positions where many are chosen than those where few are admitted. If you're a moralist and like to find morals in things, here's a second that is totally new and innovative: Things are not always as they seem. Figure 6.6 summarizes all of these facts.

Fig. 6.6 Paradoxical Percentages. When the overall totals are examined (graph at left), the percentage of males is greater than the percentage of females hired. However, when examining category by category (graph at right), the percentage of females hired is double in each category.

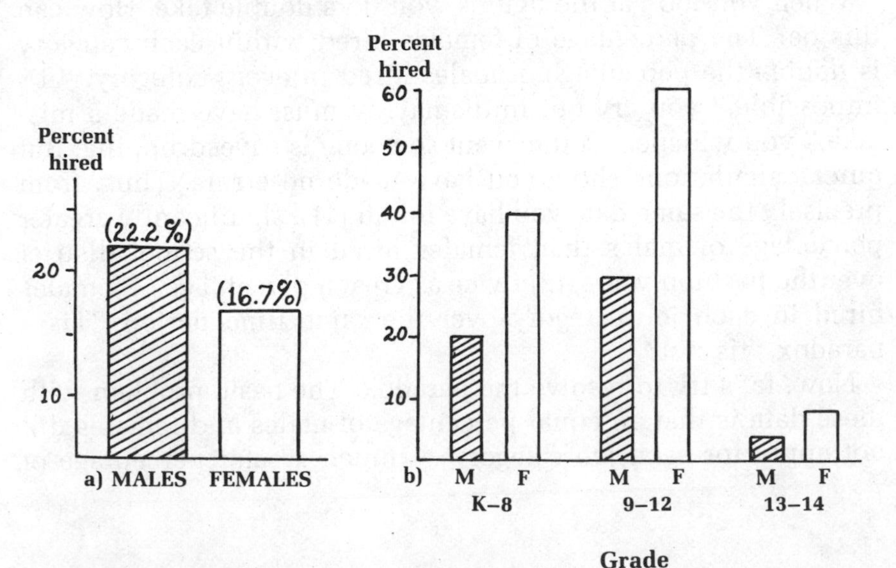

CHAPTER **7**

Percentages, Proportions, Politics, Petroleum, and Peptic Ulcers

The Percentages of Politics

A shroud of gloom had settled over the office of John Reddington Pass, candidate for mayor. It muted awareness even of the stale tobacco smoke that fouled the lungs and seared the eyes.

"So, you're saying we should throw in the towel?" It was the candidate speaking to his campaign manager. His voice was hoarse from too many speeches in too short a time, too many halls with wretched acoustics, too many cigarettes, too many promises, too many emotional highs and lows.

"In a word, yes," his manager replied.

"Damn it, Dorothy. I can't believe I hear what you're saying. You want to give up, capitulate fifteen days before the primary?" It was an effort to keep his voice on an even keel.

"I don't want to do anything of the sort. You've given it a big effort. God knows, we've all given it a big effort. But the numbers are against you, Jack."

"What numbers? Are you talking about the *Times* survey? Hell, they only asked five hundred fifty people. Five hundred fifty out of more than a hundred thousand registered voters in our city! That's less than six in a thousand. Are you telling me we should let five hundred fifty people determine the outcome of the primaries? Who are you for, Dorothy?"

"Come on, Jack. That was a cheap shot and you know it."

"I'm sorry, Dot. You know I didn't mean it the way it sounded. It's just that it's so absurd to think that a decision this important

"But the numbers are against you, Jack."

can be settled by a handful of people who happened to get tapped for a poll."

"They didn't just happen, Jack. They were very carefully selected so as to be representative of the voters in our city."

"How can five hundred fifty voters tell you anything about the will of over a hundred thousand? When I hear you statisticians talk, I think I have stepped into the world of the *Looking Glass*. No offense intended."

"None taken, Jack. I understand your feelings. You have labored hard these past few months. Everyone you talk to tells you to hang in there, you're doing a great job and you'll be our next mayor."

"Exactly! How can they all be wrong?"

"You're getting what we call a biased sample, Jack. Not many people will go out of their way to tell you they don't like you. So you don't hear much from the other side. But your friends, that's a different story. They'll tell you they like you—like your chances. That's what friends are for. So, like kings of old, you begin to think that you are universally liked and respected. How many rulers have fallen because they believed that the jackals surrounding them represented the voice of the people?"

"So now my friends are jackals? You really do love me, don't you, Dot?" It was good to see a hint of a twinkle in the corners of his eyes, bloodshot as they were. He'd make it through the depression. It would be tough, but he would make it.

"I'm your number one jackal, and I'm telling you it's time to make a strategic withdrawal."

"Do you realize what you are asking, Dorothy? Months of backbreaking campaigning. You know about that. What about the volunteers? Thousands of man-hours—"

"*Person*-hours."

"What? Ah, yes, thousands of person-hours down the drain. What about my financial supporters? Have you thought about them? What am I going to say to them? 'Sorry people—out of five hundred fifty voters, three hundred eight don't like me. This is such a crushing margin, I cannot continue the race.' Do you realize how idiotic that sounds?"

"Jack, you must face it. Your financial support has dried up. If it hadn't, I'd still give you an outside chance. But a last-minute media blitz costs money. We don't have it. We are thousands of dollars in the hole as it is. If the elections were held right now, I'd give you two chances in a thousand. Next week, with no money for radio, newspapers, television advertising—"

"Two in a thousand? Where did you dig up that figure? Remember, two hundred forty-two out of five hundred fifty said they were going to vote for me. Not two in a thousand."

"I was talking about probabilities, Jack. Here, take a look at this table." (See table 7.1.) She reached for her briefcase and withdrew a single page. "This shows the probability that you will win when we know the proportion of people in the sample who voted for you and the total number in the sample who made a choice."

"Run that by me again."

"OK. Two hundred forty-two out of five hundred fifty chose you. The proportion is .44. Go down the column headed by .44 until you come to the row where N equals five hundred fifty. This is the sample size. The number you see in the box, .002, tells you the probability that you will obtain more than fifty percent of the vote. Take a good look. Your chances are two in one thousand, Jack. It's as simple as that."

"Now wait just one cotton-picking minute. I don't see how you get from five hundred fifty people in a sample to over a hundred thousand voters in the city. You're leaping from the Colorado River up to the north rim of the Grand Canyon."

Table 7.1. Estimated probability of obtaining 50% or more of the total vote for varying obtained proportions and differing number of people in the sample. Illustration of use: A candidate receives 176 votes in a sample of 400 registered voters. This represents a proportion of 176/400 = .44. Go down the column headed by .44 until you reach the row N = 400. The entry at this point (.008) tells us that the probability is about 8 in 1,000 that the candidate will obtain at least 50% of the votes necessary to be elected. The other candidate, who received .56 of the sample vote, has about 992 chances out of 1,000 of winning.

Obtained Proportions

N	.40	.41	.42	.43	.44	.45	.46	.47	.48	.49	.50	.51	.52	.53	.54	.55	.56	.57	.58	.59	.60
100	.023	.036	.055	.081	.115	.159	.212	.274	.345	.421	.500	.579	.655	.726	.788	.841	.885	.919	.945	.964	.977
150	.007	.014	.025	.043	.071	.109	.164	.230	.312	.401	.500	.599	.688	.770	.839	.891	.929	.957	.975	.986	.993
200	.002	.005	.012	.024	.045	.079	.129	.198	.284	.390	.500	.610	.716	.802	.871	.921	.955	.976	.988	.995	.998
250	.001	.002	.006	.014	.029	.057	.102	.171	.264	.374	.500	.626	.736	.829	.898	.943	.971	.986	.994	.998	.999
300	–	.001	.003	.008	.019	.042	.082	.149	.245	.363	.500	.637	.755	.851	.918	.958	.981	.992	.997	.999	+
350	–	–	.001	.004	.012	.031	.067	.131	.227	.356	.500	.644	.773	.869	.933	.969	.988	.996	.999	+	+
400	–	–	.001	.003	.008	.023	.055	.115	.212	.345	.500	.655	.788	.885	.945	.977	.992	.997	.999	+	+
450	–	–	–	.002	.005	.017	.045	.102	.198	.337	.500	.663	.802	.898	.955	.983	.995	.998	+	+	+
500	–	–	–	.001	.004	.012	.037	.090	.187	.326	.500	.674	.813	.910	.963	.988	.996	.999	+	+	+
550	–	–	–	.001	.002	.009	.030	.079	.174	.319	.500	.681	.826	.921	.970	.991	.998	.999	+	+	+
600	–	–	–	–	.002	.007	.025	.071	.164	.312	.500	.688	.836	.929	.975	.993	.998	+	+	+	+
650	–	–	–	–	.001	.005	.021	.063	.154	.305	.500	.695	.846	.937	.979	.995	.999	+	+	+	+

700	–	–	–	–	.001	.004	.017	.057	.147	.298	.500	.702	.853	.943	.983	.996	.999	+	+	+	+
750	–	–	–	–	.001	.003	.014	.050	.136	.291	.500	.709	.864	.950	.986	.997	.999	+	+	+	+
800	–	–	–	–	–	.002	.012	.046	.129	.288	.500	.712	.871	.955	.988	.998	+	+	+	+	+
850	–	–	–	–	–	.002	.010	.040	.121	.281	.500	.719	.879	.960	.990	.998	+	+	+	+	+
900	–	–	–	–	–	.001	.008	.036	.115	.274	.500	.726	.885	.964	.992	.999	+	+	+	+	+
950	–	–	–	–	–	.001	.007	.032	.109	.268	.500	.732	.891	.968	.993	.999	+	+	+	+	+
1000	–	–	–	–	–	.001	.006	.029	.102	.264	.500	.736	.898	.971	.994	.999	+	+	+	+	+

– probability less than 0.001 (1 in a 1,000)

+ probability greater than 0.999 (999 in a 1,000)

"But we do that all the time, Jack."

"Come again?"

"We're forever making decisions based on incomplete information. When's the last time you took a bath, Jack?"

"Just like a woman to change the subject just when the going is getting a little rough. If you must know, I took a bath last night and I used a twenty-four-hour deodorant. 'Nuff said? Now let's get back to the topic at hand."

"I never left, you sexist boob."

"Thanks."

"Don't mention it. Now, when you filled the tub, did you jump right in?"

"Of course not. You think I'd risk boiling my—?"

"What did you do?"

"I dipped my foot in the water one or two times to see if the temperature was right."

"Exactly. You sampled the water and judged the whole from the part. You didn't test all forty gallons. You sampled perhaps a pint or two. What about when you're giving a speech? Do you judge its impact by looking at every face in the audience?"

"Of course not. I pick out a few faces and concentrate on their reactions."

"And if one or two go to sleep, you know you're in trouble."

"Exactly."

"Even though you have not seen every face in the audience? Again, you're judging the whole from a part. As long as that part is representative of the whole, you can keep your finger on the pulse of the audience. Talking about the pulse, have you ever thought of what medical doctors do? They take a sample of your pulse and judge whether the rate is normal or not. Same for blood pressure, blood samples, urine specimens, body temperature."

"OK. I get your point."

"The Environmental Protection Agency takes samples of air and judges whether or not stack emissions exceed acceptable levels; the Auto Emissions Test samples the exhaust gasses from your dream machine; a football coach watches you work out for a few brief practices and decides if you can make the team—"

"Enough, Dot! Enough! I get your point. Now you listen to me for a few seconds. You say we should throw in the towel based on the *Times* survey. Now answer me this. What about our own survey? That one showed me running ahead by a sixty to forty margin, and it was based on a thousand interviews. A thousand! Surely, that means more than five hundred fifty. Why, your chart doesn't even show the probability, it's so close to certain."

"Eyewash."

"Eyewash?"

"Yes, eyewash. It was conducted by your loyal followers to counter the effects of other polls. You remember what I said about bias a few moments ago? Remember the *Literary Digest* poll in 1936? The poor and the unemployed middle class were simply not represented in the survey. A large number of interviews do not compensate for a badly conducted survey. The survey of your followers was just as amateurish. Do you know what they did? They identified themselves as members of your committee and then they asked, 'Do you plan to vote for John Reddington Pass in the primaries or the other candidate?' To be honest with you, Jack, I'm not certain they didn't include themselves in the final count. The results are meaningless, utterly without a shred of validity."

"All right. I understand. But have you heard of polls being wrong? Even the scientific types conducted by *The Times*? How do you know this isn't one of those times?"

"They are not often wrong, in spite of the folklore to the contrary. [See box 7.1] We remember the few bad cases, the times that faulty sampling techniques were used. By and large, the sampling problems have been worked out. But there are contests so close that no predictions should be made. Take a look at the chart. Let us say that you obtained forty-nine percent based on a sample of one thousand. You can see that your chances are about twenty-six in one hundred. That's really too close to call. If pressed, I would have to admit that your opponent has a slight edge. But I would hedge all over the place. I know that an error of one or two percentage points could make me look like a fool."

"How about the *Times* poll? What if it made an error of one or two percentage points?"

"I have already taken that possibility into account. I asked myself, 'What if we're off by a couple of percentage points?

Box 7.1

"Excuse the interruption. I just wanted to second that point. To paraphrase Mark A., 'The good that pollsters do is oft interred with their bones; the bad lives after them.' "

"You're alluding to the *Literary Digest* poll in 1936?"

"Precisely. Someone's always bringing that one up."

"With good reason. That was a monumental boo-boo."

"Yes, but understandable. But I understand the public's paranoia about polls. After all, millions of people vote in a national election and thousands cast ballots at various levels of state, municipal, and local elections. Forecasts may be based on samples that number in the hundreds."

"Sounds more like black magic than science."

"But there's nothing mysterious about drawing inferences from samples. When we bake a cake and wish to judge whether the interior is done, we stick a toothpick in a few locations. We don't have to make a pincushion of the thing. We may use a single thermometer to judge the air temperature within a house or out-of-doors; we run a single percolator test when selecting a site for a septic field; manufacturers select a small sample of daily output in trying to estimate the percentage of defectives."

"You've made your point, Hal. Now, about that *Literary Digest* poll—"

"Please, don't interrupt. I'm getting to it. The key word to this whole process is *representative*. If the sample is selected in such a way that it is representative of the broader population to which we wish to forecast, then generalizations to that population have a high likelihood of being accurate. But if they're not—"

"The wrong candidate wins?"

"Yep, that's it in a nutshell. When errors in forecasting are made, the fault can almost always be traced to the selection of an unrepresentative (or biased) sample."

"That's what happened with the *Literary Digest*."

"Precisely. Do you remember that one? The poor *Digest* predicted that Alf Landon, the Republican golden knight, would gain a landslide victory over Roosevelt. Actually, the true outcome provided some good news and some bad news for the *Digest*. The good news was that it was right about the landslide. The bad news was that the wrong candidate won."

"How could that happen?"

"Simply enough. The sample was not representative of the voting population. The sample was drawn from telephone directories and automobile registration lists. Back in 1936, the United States was

"The good news was that it was right about the landslide. . . ."

mired in the depths of the Great Depression. Anyone with a phone or a car was almost by definition well-to-do. It so happens that in those days, the affluent tended to vote Republican, as they do today. Now, if our founding fathers had had the wisdom and foresight to define an eligible voter as someone who owns a car and has a private telephone, the *Literary Digest* would have come up smelling like roses. Instead, it sank into Chapter 11 (not of this book, but the one published by the federal government spelling out bankruptcy regulations)."

"One last question, professor. Do you think Steve Martin will ever be elected to office in Terre Haute?"

What if, by chance, we happened to get too many of your opponent's people in the poll?' Well, forty-six percent does not improve your position an awful lot. Three chances in a hundred rather than two in a thousand. I would not bet my life on it. Also, remember that the error could have been made in the opposite direction as well. Then you would have the proverbial snowball's chance."

John Pass reluctantly withdrew from the race. A thousand miles away, in another city, also undergoing the convulsions of approaching primaries, another candidate for mayor refused

Table 7.2

Candidate	Number Favoring	Proportion Favoring
A	190	.38
B	153	.31
C	90	.18
Undecided	67	.13
Total	500	1.00

to bend under the weight of adverse statistical data. Rather, both he and his campaign manager chose to fly in the face of reality and proclaim the latest poll results as clear-cut evidence of victory "within our sights." Let's listen in on a conversation between the campaign manager and a reporter from *The Gazette*.

REPORTER: What do you think of the latest survey?
MANAGER: Great. Absolutely great! My boy's on the move!
REPORTER: How do you figure that?
MANAGER: Well, you read the results, didn't you?
REPORTER: That's why I asked. I'm looking at them right now. Your boy (C) is running dead last. A is in front of him by more than two to one. (See table 7.2.) Even B is getting about 1.7 times as many nods. At the beginning, nobody gave her a ghost of a chance.
MANAGER: But that survey was taken right after the debate. Look at what my boy was doing before the debate and you can see what strides he is making. (See table 7.3.)

Table 7.3

	AFTER		BEFORE		CHANGE	
Candidate	Number Favoring	Proportion Favoring	Number Favoring	Proportion Favoring	Number Favoring	Proportion Favoring
A	380	.38	300	.30	+80	+.08
B	360	.36	380	.38	−20	−.02
C	140	.14	100	.10	+40	+.04
Undecided	120		220		−100	

REPORTER: What I see is that A came out the big winner. He gained eighty straw votes. That's twice as much as your candidate. His proportion of gain is .08 to .04. He's gaining twice as fast as C.

MANAGER: But you're missing the most important single factor. A's percentage gain is

$$\frac{380 - 300}{300} \times 100 = \frac{80}{300} \times 100 = 27\%$$

My boy's percentage gain is

$$\frac{140 - 100}{100} \times 100 = \frac{40}{100} \times 100 = 40\%$$

That's a rate of gain almost one and one half times as great as A. The old steamroller is just getting into high gear.

All of which proves that, given the freedom to select your own base, it is possible to prove almost anything you've a mind to. Back when it was fashionable to boast about growth, a community could increase from ten to forty persons in a year, and flaunt a three hundred percent increase in population!

When you are at the bottom to begin with, it does not take much of an upward move to produce an impressive percentage of gain.

Economics

There are three things in life that are certain: death, taxes, and economic indexes. If anything can be measured over time—inflation, employment, unemployment, expenditures, wages, and sin—you can be certain that an economist has devised an index to monitor its temporal fluctuations. One of the favorite devices is a *fixed-base time ratio* (also called time relative, percentage relative, and a host of other minor variations). In this type of index, you select a given year as the base year and express everything before and after it in relation to that year. Take a look at table 7.4. It shows the number of unemployed persons (in millions) between 1950 and 1979. To construct a time ratio with these data, select a year—any year—as your base year. For illustrative purposes, I arbitrarily selected 1970 as the base year. Now divide the number of unemployed found each year by 4.1 (the number of unemployed in 1970) and multiply by 100 in order to express the change as a percentage. You obtain the time ratio found in the last column of the table.

Time indexes are very useful tools. They provide a wealth of information at a glance. We know immediately that any year with an index below 100 means that the value for that year is less than the base. This is good news if we're talking about unemployment, crime statistics, automobile accidents, wholesale price index, or bubble gum sales. On the other hand, it's very bad if we're looking at gross national product, hourly wages, output per person-hour, or attendance at discos. Looking at table 7.4, we see that 1966 through 1969 were halcyon years with respect to unemployment. If you think back a moment, you'll recall that they also achieved some notoriety as years of social unrest, highlighted by protests against the Vietnam War.

But there is still something wrong with table 7.4. Note that the time ratio is based on the *number* of unemployed. Now, if the work force is changing from year to year, these data may be misleading. What if the total *number* of wage earners is increasing each year (as it is) and the percentage of unemployed remains the same? In this event, the *number* of unemployed

Table 7.4 Number of unemployed persons, in millions, and time ratio index based on a fixed year of 1970.

Year	Number Unemployed	Time Ratio	Year	Number Unemployed	Time Ratio
1960	3.9	95.1	1970	4.1	100.0
1961	4.7	114.6	1971	5.0	122.0
1962	3.9	95.1	1972	4.8	117.1
1963	4.1	100.0	1973	4.3	104.9
1964	3.8	92.7	1974	5.2	126.8
1965	3.4	82.9	1975	8.1	197.6
1966	2.9	70.7	1976	7.5	182.9
1967	3.0	73.2	1977	7.0	170.7
1968	2.8	68.3	1978	6.1	148.8
1969	2.8	68.3	1979*	6.1	148.8

* Estimated on the basis of data for the first nine months in 1979.

Table 7.5 Time ratio of the proportion of unemployed in which the percentage for any given year is obtained by dividing the number of unemployed by the total work force for that year and multiplying by 100.

Year	Percentage Unemployed	Time Ratio	Year	Percentage Unemployed	Time Ratio
1960	5.6	114.3	1970	4.9	100.0
1961	6.7	136.7	1971	5.9	120.4
1962	5.5	112.2	1972	5.6	114.3
1963	5.7	116.3	1973	4.9	100.0
1964	5.2	106.1	1974	5.6	114.3
1965	4.5	91.8	1975	8.5	173.5
1966	3.8	77.6	1976	7.7	157.1
1967	3.8	77.6	1977	7.0	142.9
1968	3.6	73.5	1978	6.0	122.4
1969	3.5	71.4	1979	5.8	118.4

must go up. Failure to realize this fact may result in a bunch of misleading statements. A prime contender is any so-called crime statistic. Politicians wishing to embarrass the incumbent are not above such proclamations as "Crime, all sorts of crime, is up all over the state. We had more murders, homicides, rapes, burglaries, and drug abuse cases last year than during any preceding year in our history. The incumbent is clearly soft on crime."

The way out of this statistical mischief is to base the time ratio on proportional or percentage increases. For example, we could modify the entries in table 7.4 to represent the proportion of the total work force that is unemployed during any given year. Similarly, adult crime data could be expressed in terms of the proportion or percentage of adults in the population. When table 7.4 is redone to express these *rates* of unemployment, we obtain the following table (table 7.5).

The difference that this change can make is dramatically demonstrated in figure 7.1. Note that the solid line representing the number of unemployed *underestimates* the *rate* of unemployment for all years before 1970, the base year. For example, looking at the solid line, we might judge 1961 to have been a sort of average year on the unemployment scene. It ranks ninth in numbers of unemployed over a twenty-year period. However, when we look at the dashed line, we see that as a matter of actual fact, the *rate* of unemployment for that year was the

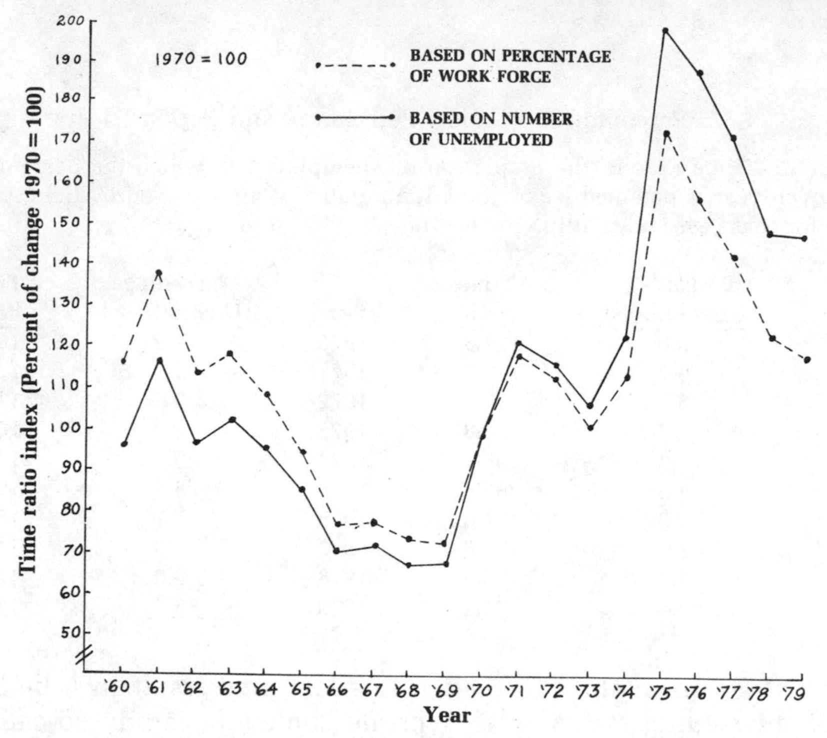

Fig. 7.1 Two Different Perspectives. Two time ratios based on the same unemployment data but differing in the impressions they convey. The solid line is based on the numbers of unemployed during each of the years. The dashed line is adjusted to reflect the percentage of the total work force that was unemployed during each year.

fourth highest over the same twenty-year period. Note that both of these are perfectly legitimate ways of showing data. Neither involves lying, deception, or chicanery in and of itself. But it's so easy to seduce the reader into drawing false conclusions, particularly if you are somewhat lax in labeling the figures and tables.

There's even more you can do, if you have a mind to deceive. So far we have said nothing about choosing the base year. By an astute selection, you can convey almost any impression you desire. I hasten to note that I'm talking about *impressions,* not reality. The truth is that no matter what base year you choose, they all tell precisely the same statistical story. One can easily be transformed into another. But they *seem* to tell different tales, and that provides ample opportunity for deception.

Take a look at table 7.5 and figure 7.2. When the very lowest year is used as the base, all other years show higher rates of unemployment. Indeed, 1975–1977 looks like the Great Depression.

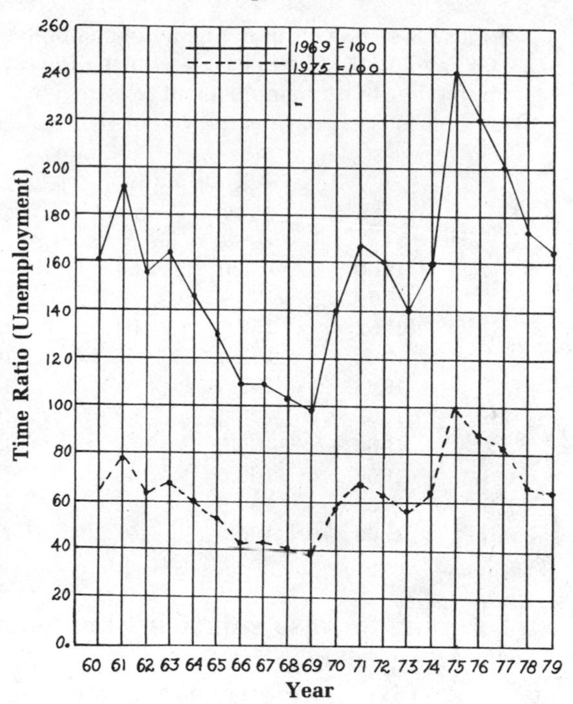

Fig. 7.2 The Same Information? Different impressions are conveyed by use of different base years when presenting graph of time ratios. When 1969 is used as base year, unemployment figures appear to be high. When 1975 is base year, unemployment figures appear to be low. In actuality, both graphs convey precisely the same statistical information.

Petroleum

Now, let's put your imagination to the supreme test. Suppose that you are a hapless oil company reeling under the continued assault of unprecedented profits. You are accused of price gouging, making windfall profits and all those other nasty things that the Seven Sisters* and their consorts never do. Now imagine that table 7.6 and figure 7.2 represent profits over the most recent twenty-day period. You have two problems: You must satisfy Congress and the American public that your profits are not all that great, yet you want to show your stockholders that they can trust the hands they put themselves into.

"Why that's easy," says Mr. $a$$et, president of Ffo-pir (pronounced foe-peer). "We take day sixteen as our base. That's the only day our profit was at a hundred percent. Every other day it was less. Why on day ten it was only 41.2. How are we ever

* Royal Dutch-Shell, British Petroleum, Mobil, Texaco, Gulf, Exxon and Standard Oil Company of California.

Table 7.6 Time ratios of unemployment rates calculated from two different base years, 1969 and 1975. The one based on the lowest year of unemployment (1969) gives the impression that unemployment levels were high in most other years. When based on the highest year of unemployment (1975), the levels at other years appear to be low.

	Year	Time Ratio 1969 Base	Time Ratio 1975 Base		Year	Time Ratio 1969 Base	Time Ratio 1975 Base
1.	1960	160.0	65.9	11.	1970	140.0	57.6
2.	1961	191.4	78.8	12.	1971	168.6	69.4
3.	1962	157.1	64.7	13.	1972	160.0	65.9
4.	1963	162.9	67.1	14.	1973	140.0	57.6
5.	1964	148.6	61.2	15.	1974	160.0	65.9
6.	1965	128.6	52.9	16.	1975	242.9	100.0
7.	1966	108.6	44.7	17.	1976	220.0	90.6
8.	1967	108.6	44.7	18.	1977	200.0	82.4
9.	1968	102.9	42.4	19.	1978	171.4	70.6
10.	1969	100.0	41.2	20.	1979	165.7	68.2

going to raise sufficient investment capital to meet our future energy needs?"

"Like taking over Montgomery Ward's, Florida Condominiums, and those other giants of energy?"

"Precisely."

"OK. So what do you say to your stockholders? Surely they can't be too happy about the dismal profit picture."

"No problem. We use day ten as our base. On that day our profit was a hundred percent. But, how sweet it is, every other day was better. Why, on days sixteen through eighteen, we more than doubled our profits over the base day. But this is not an unmixed blessing."

"How come?"

"I have inside information that OPEC is going to increase the price of crude again at ten o'clock this evening."

"So?"

"Well, that will double our profits within two days. When you use that day as our base, the profits on days sixteen through eighteen will look like huge losses. I'm afraid there'll be only one corporation large enough to justify the time and effort of acquisition at this point."

"Really? Which one?"

"The U.S.A."

CHAPTER **8**

What Average Is Most Average?

How good is your imagination? Can you imagine yourself as the head of a local labor union making preparations for a strike? You can? Good. Then imagine me as the well-fed and somewhat corpulent corporation executive preparing to do battle with you over the latest outrageous union demands. And of course you are the gaunt, emaciated representative of the downtrodden laboring class, complete with two days of chin stubble, cavernous eyes, and hollow cheeks. Of course, we know this image is a fiction. We're both pretty portly these days. In fact, as head of the labor union, you probably earn more than I do when fringe benefits are considered. And certainly your tennis club is equal to mine. But don't you sometimes long for the good old days when you were a Bolshevik agitator—a bad guy—and I was the knight in shining armor trying to protect our system from incursions of the Red tide? We'd call out the goon squads and you'd plant a few bombs. A few heads would get bashed in the process, but we'd work something out somehow. But now it's lawyers and more lawyers—countless hours of lawyer gobbledygook. Ah, the fun is gone. I sometimes think labor unions were invented by lawyers, just so they could collect a fee. I mean they invented everything else—possession of private property, marriage, divorce, cars, babies, accidents—so why not labor unions?

But I am digressing. The point I am trying to make is that the armamentarium in the union-management battles has changed radically over the years. A big part of the skirmishing now takes place between the public relations officers of both sides. And what is their main weapon? You guessed it. The Statistic. Not necessarily the statistical lie but merely the statistic deployed in such a way that the same effect is achieved.

Let's see how this works. You, the big labor boss, call on your public relations man. You say to him, "Get together some statistical proof that our laborers are underpaid. I've already reserved a full-page ad in *The Times* and I've written the ad. Now all I need is the numbers to fill in the missing spaces."

Of course, *I* don't work that way. I am honest and I always seek the truth. I call my public relations man and ask him to dig up the data on salaries, fringe benefits, and all that stuff: "Most of all I want to know what the average salary is around here. I'm sure it's one of the highest in the industry and I think we should be telling it to everybody but the stockholders."

The answer comes back promptly with the pure tones of Berlioz's *Ranz des Vaches* (I like to show off a little bit): "Our average hourly salary, exclusive of fringe benefits, is seven dollars and seventy-three cents."

"Ho, ho," I say. "That will cook their little goose-flesh behinds. Here we are fighting spiraling inflation, sacrificing, cutting our profit gains to under fifty percent of last year, and those crumbs are looking for raises. It's unpatriotic, undemocratic, and anticapitalistic."

Imagine my surprise when I open *The Times* on the following morning and see the union ad: "Peons at Wankee receive lowest wages in the industry. At $4.21 an hour, many qualify for welfare."

Now, my friend, let's step out of our adversarial roles for a moment and examine this situation. Management says that the average hourly wage is $7.73; labor claims it is $4.21. Is labor lying, management lying, or are they both lying?

The strange thing is that both may be telling the truth. In fact, they would be crazy not to tell the truth when the term *average* provides a loophole much bigger than the eye of a needle. We could easily put a camel through this one!

The term *average* is one of those ambiguous words meant to tell you something about the middle of a distribution of numbers, of scores, of wages, etc. The only trouble is that there are three commonly used measures that describe the center of a distribution. They are called *measures of central tendency* and they are defined in different ways. With many data, this difference in definition will not make a difference. With others, such as wages, the difference is as great as the disparity between

Table 8.1

Name of Employee	Hourly Wage	
110 15 2436	$ 4.00	
109 16 4134	$ 4.21	
015 62 3343	$ 4.21	Mode ($4.21)
101 45 1362	$ 4.21	
515 60 4142	$ 4.21	
612 45 3627	$ 4.81	Median ($4.51)
413 21 6561	$ 4.90	
218 35 4425	$ 5.20	
806 56 7132	$ 5.45	Mean ($7.73)
Mr. Parsons	$36.10	
Total	$77.30	

gross income (several million dollars) and net adjusted income (zero) on a rich man's tax return.

Now, let's take a peek at the hourly wages of employees at Wankee (see table 8.1). To simplify matters, we'll show only a small representative sample of the total. As a matter of actual fact, Wankee employs thousands of laborers in its East Coast plant. A large number are unskilled and obtain the lowest wages. A select few are highly skilled, possess advanced degrees, and command substantial hourly wages.

Wage and salary figures characteristically show much greater extremes at the high end than at the low end. After all, you can't earn less than zero income but there is no limit at the other end, particularly if you're Nelson Bunker Hunt or a Persian Gulf oil czar.

Three commonly used measures of central tendency are the mean, the median, and the mode. We are all familiar with the mean. When we were in school and had to calculate the arithmetic average of a set of grades, we added them together and then divided by the number of grades. So, if we got 90, 70, 80, and 80, the sum would be 320. Divided by four, this yields a mean of 80. In our example with Wankee, the sum of the hourly wage of the ten employees is $77.30. When divided by ten, a mean of $7.73 is obtained. That's where I got my "average" when I played the role of an executive of Wankee. Note one important characteristic of the mean: Every score enters into its determination. If there are extreme scores at one end of the distribution (in our example, Mr. Parsons's salary), the mean is pulled in the direction of those scores. It's like a heavyweight on one end of a teeter board. In this rather exaggerated case, the mean is a deceptive "average"—it represents no one—and I deserve to be chastised for this bit of statistical deception.

But don't think that lets you off the hook. No siree/ma'am (cross out the one that doesn't apply), you're just as big a crook as I am. You chose the mode, which is the score that occurs with the greatest frequency. You're taking advantage of the fact that Wankee employs mainly unskilled workers, all of whom earn wages that are consonant with their contribution; in other words, low on both counts. Your "average," my friend (and I use the word advisedly) represents only our lowest-paid employees.

If we were both dedicated to honesty, we would select the median as our measure of central tendency in this case. Why?

Tantalus and the Mean.

The median is the middle score. You find it merely by counting down until you locate the score that is exactly in the middle. Half of the scores are above it and half are below. In the present example, the middle score doesn't really exist, so we arbitrarily select a value that lies halfway between the fifth and sixth scores. This gives us a median hourly wage of $4.51. Note that the median provides a fairly good representation of the majority of the wage earners. Note also that unlike the mean, the median is unaffected by the extreme scores. If Mr. Parsons had earned a million dollars an hour (and revealed his true name to be Sheik Yemani), the median would have remained precisely the same. Generally speaking, the median is the measure of choice when there are extreme scores in a given direction, as with salary and wage figures. We speak of such distributions as being *skewed*. If you think of the word askew—off balance—you've got the right idea.

But the fun and games are just beginning. You've all heard of the *double standard* and mistakenly thought that it referred to male-female freedoms. The double standard consists of using two different statistics at the same time to make your point. But why am I telling you this? You know all about it. I forgot to tell you we had your phone bugged. In case you're a bit hazy, let me reproduce a bit of one of the taped transcripts to refresh your memory. The conversation is with your West Coast counterpart.

YOU: Say, Harry, it looks as if we're going to have a bit of a battle with those (expletive deleted) at Wankee over salary negotiations. Could you give me an idea of the average hourly wages out there?

HARRY: No sweat. We're getting our (expletive deleted) butts ready for negotiations here so I've got them right at my fingertips. Er . . . let's see. Ah, here they are. The average hourly salary here is four dollars! Why, I'll be a (expletive deleted)!

YOU: That's all?

HARRY: A lot of uneducated labor, you know.

YOU: Yeah, I forgot. How much is management claiming as the average?

HARRY: About six dollars and forty-five cents.

YOU: Would you mind sending me their figures? Forget about your own. It's only fair that we use figures from both sides. My figures for the East Coast and management's for the West Coast. That way, no one can accuse us of being biased. The figures also have a nice ring to them. Our guys sweat for four dollars an hour and yours get about six forty-five. By God, we're being screwed.

HARRY: I'll send you their figures under one condition.

YOU: OK.

HARRY: Yes—You send me the figures that your management has compiled. What's good for the goose is good for the gander.

YOU: (expletives deleted)!

End Transcript

Do You Mean We Shouldn't Use the Mean?

Now don't get the wrong idea from our previous example. Most of the time the mean is a pretty good measure of central tendency. It's just that when the distribution is badly skewed, the mean simply does not tell you much about the most typical or representative scores. But even in those cases, the mean provides information not available from the other two measures. For example, what if you wanted to know how much Wankee pays its ten employees each hour and the only information you have is that the mean is $7.73, the median is $4.51, and the mode is $4.21. Here the fact that the mean is based on arithmetic processes—adding, subtracting, multiplying, dividing—makes our problem a lead-pipe cinch.

Remember:

$$\text{Mean} = \frac{\text{Sum of scores}}{\text{Number of scores}}$$

By the simplest of algebraic manipulations, we can find the unknown total:

$$\text{Sum of scores} = \text{Mean} \times \text{number of scores}$$

$$\text{Sum of scores} = \$7.73 \times 10 = \$77.30$$

This mathematical characteristic of the mean makes it a very valuable statistic, in both descriptive and inferential statistics. When you add to this the fact that many of the things we measure in this world tend to distribute themselves in a symmetrical bell-shaped fashion, it really doesn't make much difference which measure you use—the mean, the median, or the mode. They are all the same. You can see this more clearly in figure 8.1.

While we're on the topic of averages, let me get one of my favorite gripes off my chest. To be most blunt about it, the average person bugs me. That's right. I hate the average person. Before you get me wrong, let me explain what I mean. Somebody starts out with a few statistical facts. The average (mean) IQ of the general population is 100. The average (mean) height of males is five feet eight inches; of females, five feet four inches

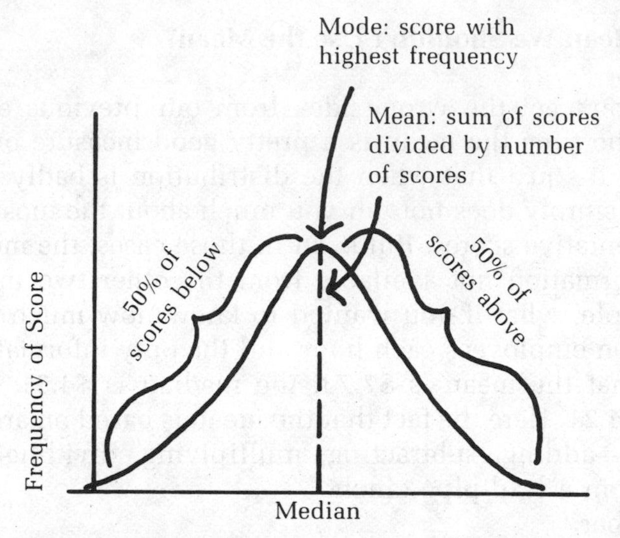

Fig. 8.1 Symmetrical Distribution. In symmetrical bell-shaped distributions, the mean, median, and mode are all the same.

(I just made those figures up. I forget what they are and it really doesn't make any difference for the point I am making). The next thing you know a subtle semantic transmogrification takes place and we hear, "The average person has an IQ of one hundred;" "The average man is five feet eight inches tall;" and so on, *ad absurdum, ad nauseam*. I don't have any idea what sort of creature the average person is, but of one thing I am sure. Having one attribute that is average does not make the individual average. I know a clod who is eight feet two inches tall, weighs 400 pounds, giggles, and has six toes on his third foot. But he has an average IQ. Does this make him average? Oh, how I hate the average person. For that matter, I hate the average anything.

Straining under the Weighted Mean

"I'll tell you what I'm gonna do," says the carpet salesman with the slicked down hair and the Fu-Manchu mustache. "I'm gonna make you an offer you can't refuse."

"Oh?"

"You want 3,000 square yards for that apartment house you're fixing up, right?"

Table 8.2

X Grade of Carpet	W Quantity in sq. ft.	WX
9	1650	14,850
11	500	5,500
12	350	4,200
13	250	3,250
15	250	3,750
	Sum: 3,000	31,550

"Right."

"You've selected five grades of carpeting . . . \$9, \$11, \$12, \$13, and \$15. That's an average of \$12 a yard. Multiply that by 3,000 yards and we get, let's see . . . \$36,000. I'll give you the whole lot for \$32,000. That's an 11 percent discount over our already low prices."

How does that sound to you? Is it really a good deal? It depends. Our slick friend has calculated a very deceptive mean. He just took the price tags and averaged them without regard to the quantities of each grade of carpeting you are buying. Here's the rub. The sum of the prices of these grades divided by five gives us the mean *only* when we buy equal quantities of each. But what if this isn't the case? Then we must calculate the weighted mean. Let's say you wanted to purchase the quantities of carpeting shown in the second column of table 8.2.

To find the mean here, you must multiply the price of each grade by the quantity ordered (W), add them together, and divide by the sum of the weights. In other words:

$$\text{Weighted mean} = \frac{\text{Sum } WX}{\text{Sum } W} = \frac{9 \times 1650 + 11 \times 500 + 12 \times 350 + 13 \times 250 + 15 \times 250}{1650 + 500 + 350 + 250 + 250} = \$10.52$$

Aha. The actual mean is \$10.52, rather than the \$12.00 represented by your carpet-dealing carpetbagger friend. Your initial cost estimate should have been $31,550. Four hundred and fifty dollars less than the magnanimous offer. Some deal! Chalk up the extra 450 bucks as the price of ignorance.

Weighted averages are used and abused much more than you might think in the shenanigans of daily living. If you bought 100 shares of stock at $5.00 a share, 50 at $6.00, and 10 at 7.00, your break-even price is not $6.00. It is:

$$\frac{100 \times 5 + 50 \times 6 + 10 \times 7}{100 + 50 + 10} = \frac{870}{160} = \$5.44$$

Similarly, if your dream machine guzzles an average of 20 mpg on a 300 mile trip, 23 mpg over 800 miles, and 25 mpg over 1000 miles, your average miles per gallon is:

$$\frac{20 \times 300 + 23 \times 800 + 25 \times 1000}{300 + 800 + 1000} = \frac{49{,}400}{21{,}000} = 23.52$$

CHAPTER **9**

A Standard Deviation Is Not Your Run-of-the-Mill Perversion

Take a look at that cartoon again. It is telling an important story. We tend to get terribly hung up on what's average. We forget that few things are average on any given characteristic that we wish to measure. How many people do you know who are of average height, weight, and IQ, who come from average families in average crime areas, and who own average houses with average-sized mortgages on average-sized plots? The truth of the matter is that the three common measures of central tendency

"I don't care if he's a little large as long as he's average."

are statistical abstractions. Proportionally, very few members of a given population achieve an "average" value on any measure that we wish to examine. I think we can assure ourselves that very few parents have achieved that wondrous and advanced state of planning in which they have exactly the number of children, 1.33, that statistical tables tell us is the average (mean). The truth of the matter is that most measures in the real world are either below average or above average. Failure to recognize this ineluctable fact of statistical life leads to that delightful state of affairs known as a *WOW statement*. What's a WOW statement?

Do you know that half of the people in the United States are above average in weight? WOW!

Do you know that half of our children are below average in IQ? WOW!

Do you know that half of our people are below average in emotional balance? WOW!

Do you know that half of the Japanese are above average in emotional balance? WOW!

Do you know that half of our cars are below average in safety? WOW!

Do you know that half of foreign cars are above average in safety? WOW!

Prepare yourself for the next one. It is heart-rending, a tragedy that science must do something to correct.

Do you know that half of the people in this country die before they reach average age? WOW!

And that half the people in other civilized nations die at or above average age? WOW! WOW! WOW!

It's just as we always suspected. American medicine is like a freight train going downhill without brakes (although the costs are going in the opposite direction with jet propulsion). The reason? An above average number of shows on television are about doctors, nurses, hospitals and the shenanigans that take place underneath those white coats. With all of those doctors on TV, how many are left for us?

Failure to recognize the abstract nature of "average" can also lead to a lot of mischief. Many parents, reading the norms compiled by child psychologists, grimly resolve that, by hook or by crook, their children will not be below average on anything. (And I mean *anything*.) Consideration of this sort leads to another class of shockers, which I refer to as *OY statements*.

Do you know that the average child walks with support at forty-eight weeks of age? Your Johnny is a year old and he is still grounded on his tush. OY!

Do you know that the average child speaks her/his first word at thirteen months of age? Mary is fourteen months old and still makes bubbles when she says "goo." OY!

Do you know that the average child stops making messy-messy by thirty months of age? Your Theosopholus is thirty-four months old and he still stinks to high heaven. Damn brat!

The truth of the matter is that this world of ours is wonderfully varied. Thank the Creator that we are not all average everything or anything. Of course if we were, it would simplify things. Imagine telephoning the local clothing store and saying, "Send me an average pair of pants or an average dress," or telling your friendly automobile dealer, "Send me an average lemon." Convenient, yes, but what a crushing bore!

But we are different from one another. Things are different from one another. We are even told that no two snowflakes are ever alike (I'd like to see how anybody would go about proving *that*). Because of the wide variability of all things measured, it is clear that measures of central tendency have only limited value in describing the totality of events that interest us. Clearly some companion to central tendency—a descriptive measure of the spread of scores about central tendency—is desired. "Why?" you ask.

Imagine that you are a home builder and you know that the average number of people per household is 2.81. (That's what it was in March 1978; note that again we have the statistical abstraction. I am willing to give considerable odds that there are few families of median size.) You are planning to put in a housing development. How many bedrooms should you provide? Our measure of central tendency does not help very much. What you need is some information about how family size is distributed, spread out, or dispersed throughout the population. What proportions of families are two person, three person, four person, five person, and so on? Data such as the following (which

I took from the *American Almanac of 1976: The Statistical Abstract of the U.S.*) are far more useful.

Size of Family	Number (Thousands)	Percent
2 persons	20,592	37.4
3 persons	11,673	21.2
4 persons	10,789	19.6
5 persons	6,386	11.6
6 persons	3,021	5.5
7 or more persons	2,593	4.7

This information tells us that were we to build all housing units to accommodate two- and three-person families, we would be neglecting about 41 percent of the population. That's a lot of people and a lot of houses.

Or take another example. Your little twelve-year-old Penelope has just returned from school.

> YOU: How was school today?
> PENELOPE: You know, fine. You know.
> YOU: What did you do?
> PENELOPE: Nothing, you know.
> YOU: Nothing?
> PENELOPE: You know, nothing much. Took a test, you know. You know, one of those tests of scholastic achievement.
> YOU: A test of achievement? Well, how did you do?
> PENELOPE: You know, OK, I guess. You know.
> YOU: You guess?
> PENELOPE: Yeah, you know. I guess so. I got, you know, fifty-five right. You know.

Now at this point do you congratulate Penelope for doing so well, mildly chastise her for not doing better, or excoriate her for doing so poorly? (Or put her over your knee for all of those ridiculous "you knows"!) None of these, of course. You simply have no standard by which to judge a score of fifty-five correctly. So you probe further.

> YOU: What was the class average?
> PENELOPE: Dear parent, you should know that the word average is imprecise and ambiguous [snot-nosed kid!] There

are three measures of central tendency, you know, the mean, median, and mode. Which would you like to know?

Now it is your turn to show off your erudition.

YOU: Is the distribution of scores symmetrical or skewed?

PENELOPE: It's a, you know, standard educational test. You know, that means it's symmetrical. You know. It's, you know, bell shaped.

YOU: Well, if it's bell shaped, what difference does it make which measure of central tendency you tell me? They're all the same.

PENELOPE: I didn't, you know, know you knew. The mean for the test is fifty.

What do you know at this point? Not really very much. You know that Penny scored higher than the mean. Since the mean and median are identical, you also know that more than 50 percent of the people taking the test scored lower than your daughter. So the question "How good is a score of fifty-five?" must await additional information about how the scores are dispersed around the mean. So you continue the grilling.

YOU: What is the highest score?
PENELOPE: Seventy-five.
YOU: And the lowest score?
PENELOPE: You know, twenty-five.

Now, you know, we know the range of all possible scores (75 − 25 = 50). This tells us something about the dispersion of scores, although not very much. We know that Penelope's score, albeit above the mean, is not near the top. But it's closer to the top than if the range had been 0–100. On the other hand, if the range had been 45–55 . . . oh, blessed nirvana! All of this is made visual in figure 9.1.

The range is not usually expressed in terms of the highest and lowest values actually achieved by a group in which we are interested. It is entirely possible that in Penny's class the highest obtained score was below seventy-five; perhaps even as low as fifty-five. If this were the case, Penny's score of fifty-five would begin to take on a different appearance.

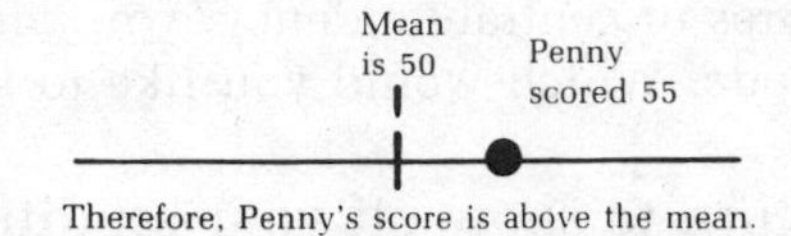

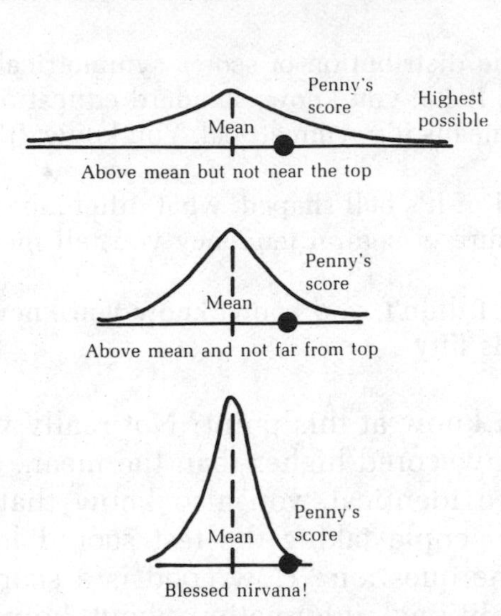

The value of the range is further complicated by the fact that many of the dimensions that pique our curiosity contain bottom scores but the upper end is wide open. A few very extreme cases will distort our perception of the range within which most values fall.

Consider the following.

Income. When this drops down to zero, it's at rock bottom. You can't go lower. But what's the top? Nowadays, I guess the answer depends on how much oil you're sitting on and how fast you can squeeze it out. Certainly the energy crisis is bound to create more than a few billionaires in scattered parts of the globe. It will also leave in its wake many more than a few paupers.

Family size. The number of children in a family cannot fall below zero, but the upper end is far less precise and limited. The *Guinness Book of World Records* reports a substantiated case in which a Russian woman, Mrs. Fyoder Vassilet, gave birth to 69 children. Excuse me for saying this, but I can't resist the temptation. Since Mrs. Vassilet lived a total of fifty-seven years, she averaged 1.2 children per year from the nursery to beyond Medicaid. Stated another way, she averaged one tenth of a child every month of her life.

These are just a couple of examples. I could cite many more. While athletes of the world are stretching for that 3:49 mile, the 12-minute mile still eludes my plodding legs. I am also getting rather portly with increased age, but thank heaven I'll never threaten the world record of 1,069 pounds. No piano-case burials for me, thank you.

What's the point of all this? Simply that the range, while useful in depicting the total spread of values on any given characteristic, represents only the two most extreme values. Knowing a measure of central tendency and the range does not allow us to grasp the dynamics of the distribution of scores between the two extremes. After all, knowing that the heaviest human being weighed over half a ton and the lightest adult balanced the scale at the weight of a good-sized trout (4.7 pounds) does not tell us an awful lot about the range between which *most* human weights fall. The range is simply the victim of the two most extreme values.

There is a measure of dispersion or variability that is absolutely ideal so long as one condition is met—the scores must be distributed in a bell-shaped fashion, known as the *normal distribution*. You will recall that this is the type of distribution in which all three measures of central tendency are identical. The ideal measure of dispersion, or spread, under these circumstances is the *standard deviation*. It is ideal for many reasons, but one stands out above all the others: Once you know the mean and standard deviation of a distribution of scores, you know all that is necessary to interpret *any* scores and you can reconstruct the entire original distribution of scores, if you like. Let's examine that statement a little more closely.

Imagine that we have just established radio contact with a civilization out there in space somewhere. (This could well happen within the lifetime of most of us. With billions of stars in billions of galaxies, it seems inconceivable that earth is alone endowed with intelligent life. Indeed, it is probable that there are some civilizations out there that are as advanced over us, in knowledge and technology, as we are over the Neanderthals. If this is so, it is unquestionable that some have been sending out radio signals—perhaps for thousands of years—in hopes of

"I'm your friendly standard deviation for October, Sir."

intercepting life elsewhere. What has changed for us is the fact that within the past few years we earthlings have developed for the first time the capacity to receive coherent signals from outer space. Moreover, our capabilities are improving almost daily.) So let us assume that we have established contact with a civilization five light years away. This would mean that we could complete a two-way communication every ten years. Perhaps the first twenty years would be spent trying to overcome the language barrier, with the language of mathematics providing the breakthrough. After that, we would want to begin sharing personal information about each other. What do we look like? What is our means of reproduction? If it is sexual, how do the sexes differ in overall appearance and in all specifics such as height and weight? At this point, the magnificence of statistical analysis would emerge. All that would be required to describe most human characteristics would be two bits of information on each: the mean and the standard deviation. Since the civilization living five light years away would almost surely have independently discovered the mean and standard deviation, imagine the wealth of statistical information we could convey by a series of short radio bursts.

Height
Male: Mean/Standard deviation/
Female: Mean/Standard deviation/
Practice of premarital interdigitation
Male: Mean time/Standard deviation/
Female: Mean time/Standard deviation/

Putting aside extraterrestrial communications, we can see that we humans could even communicate much information to each other if more of us understood the language of statistics. Let's look at a few examples that illustrate the use of the mean and the standard deviation when dealing with normally distributed scores. The three children of your Aunt Mathilda—Mary, David, and Laurie—have just taken standard educational tests in school. Their scores on each of the tests were as follows:

Spare Parts Identification Test: Mary's score = 80
Clerical Aptitude and Nomenclature: David's score = 70
Car Aptitude and Nomenclature Test: Laurie's score = 650

Aunt Mathilda knows that you are very smart so she singles you out for advice. "What shall I do with these scores?" she asks pleadingly.

Successfully fighting off the temptation to give her a smart reply, you answer instead, "Go to the *Mental Measurement Yearbook* and find the mean and standard deviation of each of these tests."

"But how will that help me?"

"Get me the information I requested and I shall explain," you whisper with a conspiratorial air. "The interpretation of test scores involves a closely guarded secret code that only a gifted few have managed to crack."

"Oh, how thrilling!"

Breathlessly, she returns a few hours later. Voice quivering with excitement, she reveals all: "The mean on SPIT is a hundred, with a standard deviation of ten; David's CAN has a mean of fifty with a standard deviation of ten; and the mean of Laurie's CANT is five hundred with a standard deviation of one hundred."

"The first thing we must do now," you say, resuming your secretive air, "is to arrange each score with its accompanying mean so that we can make certain comparisons. So this is what we have."

Individual	**Test**	**Score**	**Mean**	**Difference**
Mary	SPIT	80	100	−20
David	CAN	70	50	+20
Laurie	CANT	650	500	+150

"Oh? Huh?"

"Now, if we subtract the mean from each score, we find out two things."

"What things?"

"For one, whether a score is above or below the mean. A minus score indicates below the mean. Second, we learn *how far* the score is above or below the mean."

"Oh, I see. Mary's difference or deviation score is minus twenty so she is twenty points below the mean. Laurie is one hundred fifty points above the mean. That means she did the best."

"Not necessarily. That's where the standard deviation comes in."

"Oh, that thing again. I don't have the foggiest notion what it is, but it sounds kind of naughty."

"Naughty only if you don't understand it and it keeps you in the dark."

"Anything that keeps you in the dark can't be all bad."

"Actually, it's a lot like a car. You don't have to know how it's put together in order to use it. Now, listen closely. I am about to reveal the most intimate secret of the standard deviation. The first thing you do is divide each of those difference scores—also called deviations from the mean—by its corresponding standard deviation."

"Wait a minute, you've lost me."

"OK. Step by step. Mary got an SPIT score of eighty. Since the mean SPIT is one hundred, her difference score is minus twenty. Dividing this by the standard deviation of SPIT yields a final score of minus two. This final score is called a z-score. Now David's CAN."

"Rein up, cowboy, you're losing me."

"O.K. Let's go over the logic of transforming to z-scores. You have the scores of your children on three different tests. You know the mean and standard deviation of each of these tests. Right?"

"Right."

"You can subtract the mean from each score to determine how far it is above or below the mean. Right?"

"Right on."

"Now you can divide each difference by its standard deviation. In doing so, you express the difference in standard deviation units."

"You just lost me."

"OK, try this one. Kareem Abdul Jabbar is about eighty-seven inches tall. How tall is cousin Mary?"

"Fifty-one inches."

"How much taller than Mary is Kareem?"

"Well, let's see. That would be eighty-seven minus fifty-one is thirty-six."

"Good. Now, how do you express this difference in terms of feet?"

"There are twelve inches in a foot . . . divide by twelve. So he's three feet taller than Mary."

"Good, the difference between Kareem and Mary, expressed in feet, is three. Similarly, the difference between Mary's score and the mean, expressed in standard deviation units is minus two. So you see, it's the same—to express deviations in inches in terms of feet, you divide by twelve; to express deviations from the mean in terms of standard deviation units, you divide by the standard deviation."

"That sounds simple enough."

"Now, let's do the same with David's and Laurie's scores. Go ahead, you try it. Remember, when we subtract the mean from a score and divide by its standard deviation, the resulting score is called a z-score."

"Please stop interrupting me. You're spoiling my concentration. Let's see, David's score is twenty points above the mean. The standard deviation is ten. Therefore, his z-score is twenty divided by ten, equals plus two."

"Beautiful. Laurie's?"

"She's one hundred fifty points above the mean. The standard deviation is one hundred. Her z-score is 1.50."

"Again, I say beautiful."

"So again I say, what do I do with it?"

Gnashing your teeth again to avoid a smart and expletive-deleted reply, you answer, "All you need now is a table that gives you the percentage of cases below a given score."

"Come again?"

"Take a look at this table. [See table 9.1.] It's a simplified version but it will serve to make my point. In column A are the z-scores. Column B shows the percentage of cases that obtained lower z-scores. Mary got a z-score of minus two, right?"

"Right."

"Looking at column B adjacent to the z-score of minus two, we find that the percentage of cases with lower z-scores is two. We call this percentage a percentile rank. Mary's percentile rank is two, which means that she scored higher than two percent of the people taking this test. If you look at column C, you'll note that ninety-eight percent scored higher than she."

"She didn't do so well, did she?"

"No, but maybe it's not of earthshaking importance that she does well in SPIT. When glancing at her record a few months

Table 9.1 Percentage of scores above and below a given z. Column C provides the percentile ranks of z-scores for normally distributed variables.

A	B	C	A	B	C
z	Percentage of Cases Below	Percentage of Cases Above	z	Percentage of Cases Below	Percentage of Cases Above
−2.2	1	99	0.1	54	46
−2.1	2	98	0.2	58	42
−2.0	2	98	0.3	62	38
−1.9	3	97	0.4	66	34
−1.8	4	96	0.5	69	31
−1.7	4	96	0.6	73	27
−1.6	5	95	0.7	76	24
−1.5	7	93	0.8	79	21
−1.4	8	92	0.9	82	18
−1.3	9	91	1.0	84	16
−1.2	12	88	1.1	86	14
−1.1	14	86	1.2	88	12
−1.0	16	84	1.3	91	9
−0.9	18	82	1.4	92	8
−0.8	21	79	1.5	93	7
−0.7	24	76	1.6	95	5
−0.6	27	73	1.7	96	4
−0.5	31	69	1.8	96	4
−0.4	34	66	1.9	97	3
−0.3	38	62	2.0	98	2
−0.2	42	58	2.1	98	2
−0.1	46	54	2.2	99	1
0.00	50	50			

ago, I noticed that she scored one hundred thirty-five on a standard IQ test."

"That's good?"

"Well, you figure it out. The mean of this IQ test is one hundred and the standard deviation is sixteen."

"Her deviation score is one hundred thirty-five minus one hundred. That's thirty-five."

"Right."

"Now, you divide by the standard deviation. Let's see. Her z-score is about plus 2.2. Hey, that means her percentile rank is ninety-nine."

"And that means?"

"Ninety-nine percent of the kids in her age group got scores lower than she."

"How about David and Laurie?"

"Well, David's z-score is plus two. That means his percentile rank is . . . let's see . . . ninety-eight. And Laurie's is ninety-three. Some kids I've got."

"Now on the three tests—SPIT, CAN, and CANT—who did the best?"

"Now wait a minute. You can't compare apples and oranges."

"Can't you?"

"Of course not. On second thought . . . well, of course. David's z-score on CAN was the highest and Mary's SPIT z-score the lowest. In terms of relative performance on each test, David did the best, Laurie next, and Mary's SPIT was outrageous. You *can* compare apples and oranges. Yikes!"

"Well, auntie dear, did you learn anything today?"

"You're damned right, I did. Nobody's gonna snow me into believing test scores are magical numbers that can be interpreted only by the chosen few."

"Anything else?"

"Statistics can be fun."

"You win the brass ring for that one, auntie."

Now, a few words of warning. Don't treat the test score as fixed and unchangeable. We have a nasty habit of pigeonholing people on the basis of test scores. "We have found your score for ever and ever. Now we'll put you in your little box and let us hear no more from you." This is truly the tyranny of testing! In truth, a test score tells us something about your level of performance at the time you took the test. We hope it reflects some of your long-term and prevailing characteristics. Otherwise there is not much sense in submitting to testing. However, we must never lose sight of the fact that there are many transient forces, external as well as internal, that continuously assail us. Sometimes a score may be the offspring of these temporary affairs rather than the progeny of enduring internal relationships.

Another thing. Read very carefully the description of the standardization group. This is the group of individuals on whom all the test norms or standards are based. This is also the reference group against which all scores are compared. Most tests used in this country are standardized on middle-class white

Box 9.1 So You Want to Interpret a Test Score?

Hal Fast has just provided me with the step-by-step procedures for taking the mystery out of the interpretation of test scores on standard psychological and educational tests.

1. Determine the mean and the standard deviation of the test. Sometimes different means and standard deviations are given for different age groups. Be sure to find these two measures for the age group in which you are interested. Sources of this information are the administration booklet for the particular test and the *Buros Mental Measurement Yearbook.* Since the administration booklets are not usually available to nonprofessionals, the *Mental Measurement Yearbook* is your best bet. If it is not found in your local library, it is almost certain to be in the collection of your nearest college or university library.

2. Transform the score you are interested in interpreting to a z-score using the following formula:

$$z = \text{score} = \frac{\text{Score} - \text{mean}}{\text{Standard deviation}}$$

If you are interested in interpreting a score of forty and know that the mean and standard deviation are thirty and 9, respectively, you would have

$$z = \frac{40 - 30}{9} = \frac{10}{9} = 1.1.$$

3. Look up a positive value of 1.1 under column B of table 9.3. Here we find an entry of eighty-six. This means that 86 percent of a comparison group with which this score is being compared obtained scores lower than eighty-six. Only 14 percent (column C) scored higher.

There it is. It's as easy as that.

individuals coming from English-speaking families. Also, the standardization groups for some tests are highly specialized, e.g., students graduating from four-year colleges. Obviously, the interpretation of a score must take the characteristic of the standardization group into account. A percentile rank of fifty on a mathematics aptitude test is not much to write home about if the standardization group is made up of eight-year-olds (unless you're younger than eight) but it's a cause for rejoicing if the standardization group consists of Ph.D.'s in mathematics!

CHAPTER **10**

We've Been Going Together for a Long Time But You Still Don't Turn Me On

I have a dear friend who is gifted with a marvelous sense of humor and perfect timing to go along with it. At one time she was rather pleasingly plump and about five feet four inches tall. She no longer is (plump, that is—she's still five feet four inches tall) since going on one of those fad diets. She likes to tell the story about one of her visits to the family doctor for her semi-annual checkup.

"Get on the scale, dear. Let's see if you're being a good girl. Hm, one hundred forty-four pounds. And how tall are you?"

"Six foot two, doctor."

"Six foot two? You're not anywhere near six foot two!"

"You want I should be overweight?"

This story can be used to illustrate many points. However, I have included it to explain something that we all have observed in everyday life—scores or values of two variables often go together. In the parlance of statistics, they are correlated. Tall people tend to be heavier than short people. Conversely, short people are usually lighter than tall people. And those in between on height are usually in between on weight. Adults high on intelligence are usually high on income, credit ratings, reading ability, and, for all I know, maybe even fat!

The fact that variables go together—or are correlated—provides the scientist with a powerful tool for predicting one event from knowledge of the other. If we know such things as barometric pressure, relative humidity, and/or wind direction, we can predict weather with an accuracy far greater than we usually

Box 10.1 Flush Me a Commercial

If you were to conduct a house-to-house survey of television viewing, the educational channels would seem to dominate nighttime viewing. And how many homes do you imagine would admit to subscribing to such magazines as *Playboy*, *Playgirl*, *Penthouse*, or *Oui*? Now if we could find indirect measures—measures correlated with what we want to assess but sufficiently different that the respondent's defenses are not aroused—we'd be able to get at the truth without offending anyone. One of my favorite examples of such an indirect measure is the *flushometer*. What's the flushometer?

Some years ago a water district on Long Island puzzled over the fact that its demand for water during the evening hours was punctuated by short bursts of enormous activity, followed by long periods of quiescence. At first the water company was at a complete loss as to what forces were orchestrating and synchronizing these visits to the family john. The picture that these nightly activities conjured up was eerie. Was some UFO out there in space exercising mind control and directing people to march robotlike to the throne? A possibility, however remote.

Then, some bright-eyed individual got the idea that the toilet flushes might coincide with the commercial breaks on television

It may seem strange, but the key to water use control is the Neilsen ratings.

programs. He followed up his hunch by recording the exact amount of water demanded during specific time periods in the evening and subsequently determining when the commercial breaks took place on the television programs being viewed during these hours. The results were most astounding. It was found that the greatest number of flushes occurred during the commercial breaks of those programs that ranked highest on the Nielsen ratings. In fact, over the first ten ratings, the correlation was absolutely perfect. The number one program, which was "I Love Lucy," received the greatest surge of flush power during its commercial break, and the next nine in order showed decreasing amounts of water consumed during their corresponding commercial breaks. In this one instance, at least, reasoning from correlation to causation would seem to be fairly straightforward and direct. Few would argue that the flushing of toilets was causing the commercials to go on, but one would probably be reasonably correct in hypothesizing that the commercials were sending people scurrying to the bathroom.

Imagine the flush power if all TV programs synchronized their commercial breaks! End of the energy crisis.

admit when we roast TV weatherpersons over the coals. If we know a student's high school academic record we can make a pretty good guess about his or her performance in college. If we work in a mail-order house and we weigh the incoming mail each day, we can make a reasonably accurate prediction concerning the number of orders that will be contained therein. By making certain assumptions concerning the number of cars that will be on the road, the AAA can arrive at a frighteningly accurate guesstimate concerning the number of traffic fatalities that will occur on a given holiday weekend.

The way in which two variables are related can be shown graphically, as in figure 10.1. This figure shows the expected range of weights for small-bodied women (as judged by glove size) at varying heights. Note that a band of weights is given for each height. This is because height and weight (like most other pairs of variables that scientists study) are not perfectly related. Some tall people are lightweights, while occasionally one finds a short person who is as wide as he is tall. Such aberrations are sufficiently rare that they are celebrated in song. "Mr. Five by Five" was a hit parade tune that I recall from a youth misspent in listening to swing (as my generation called it) or noise (as

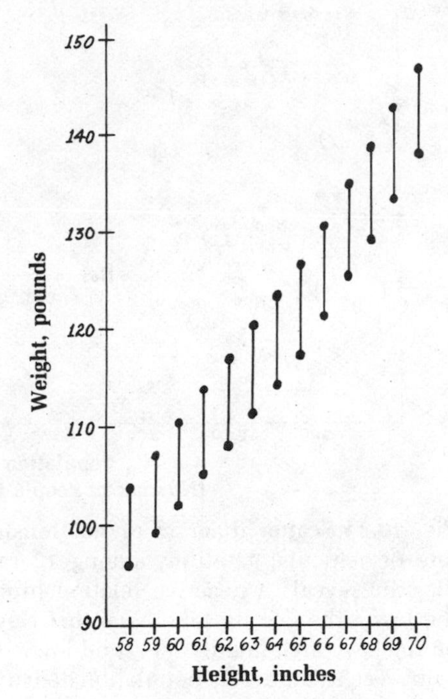

Fig. 10.1 Graph showing the expected range of weights for small-bodied women at varying heights.

perceived by the jaundiced ear of my parents' generation). Now if you really want to hear noise, listen to the stuff our kids are deafening themselves with nowadays. But I am wandering.

The point is that many events in the world about us vary together or, in the language of statistics, are correlated. The correlation may be direct (positive) or inverse (negative). The weight diagram given earlier (figure 10.1) is an example of a positive correlation. We saw that physical size and weight tend to go together. The relationship is called direct or positive because people at the low end of one scale (height) tend to be at the low end of the other (weight), those in the middle of one scale are generally in the middle of the other, and finally, those at the high end of one scale are often at the high end of the other.

Before looking at figure 10.2, let me raise a few questions. Answer them as well as you can but don't be embarrassed if you make an occasional boo boo. States with high population densities certainly suffer a higher rate of traffic fatalities compared to those with plenty of elbow room, right? New Jersey, where they pack 'em in like sardines (about 950 per square

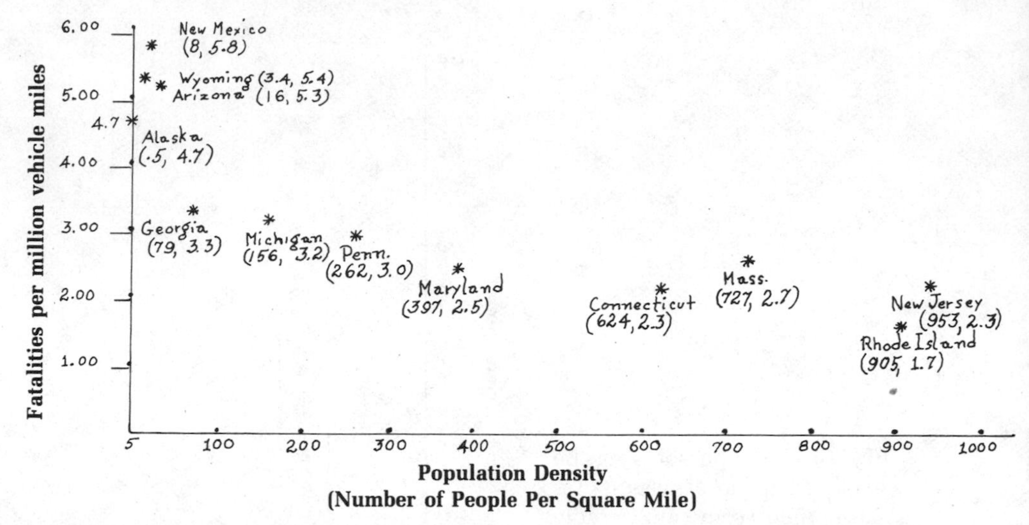

Fig. 10.2 Scatter diagram of relationship between population density and rate of vehicular fatalities among 12 randomly selected states. The scatter diagram reveals a negative relationship: The greater the population density the lesser the fatality rate. Note that New Mexico, Arizona, Alaska, and Wyoming have high fatality rates and low population densities. The two numbers below each state are its population density and vehicular fatalities respectively.

mile), must be pretty inured to carnage on the highway, right? And Alaska, with a population density so small (one half a person per square mile . . . strange looking creatures) that they have sex by telegraph, must impregnate the days of paramedics with yawning boredom, right? I hate to shatter illusions in this wonderland of statistics but the answer is wrong on all counts.

When we wish to display graphically the relationship between two variables (in this case, between population density and rates of traffic fatalities) we frequently subpoena the scatter diagram or scattergram. What is a scatter diagram? In a word, it is a means of showing paired measurements on two variables. The values of one variable are shown on the bottom or horizontal axis, and those of the second variable are shown on the vertical axis. Figure 10.2 shows a scatter diagram relating population density to the rate of vehicular fatalities for twelve randomly selected states. The relationship is definitely negative. States with low densities suffer higher rates of vehicular manslaughter. Perhaps the wide open spaces invite greater recklessness or resistance to established authority. I don't know. I do know that my own beautiful, cruel and perfidious adopted state of Arizona is, as of this writing, in danger of losing millions of bucks in federal highway funds. It seems that sixty-two percent of vehicles that have been monitored travel in excess of 60 miles per hour. Speeding is the macho thing for cowpokes to do. So it has come to this. America's love affair with the dream machine may be nothing more than a disguised form of death wish.

Now that we are familiar with the scattergram, let's take a look at figure 10.3. Again the automobile figures prominently. This time it is a scatter diagram of the relationship between the weight class of an automobile and the number of miles per gallon of each weight class. This diagram makes vivid the extent to which the large car is a gasoline guzzler. Note how the big bullies of the road are feeble onlookers in the mpg derby. The lightest cars obtain, on the average, about twenty-four miles per gallon whereas the 2 1/4 ton monsters do well if they chug along for nine miles on each gallon of the liquid gold. On ten gallons of the stuff a small car can go from New York to Boston and still have enough left over to visit Copley Square. But the guy with the big car—forget it. He'll be digging in the wallet for a refill somewhere between New Haven and Hartford, Connecticut.

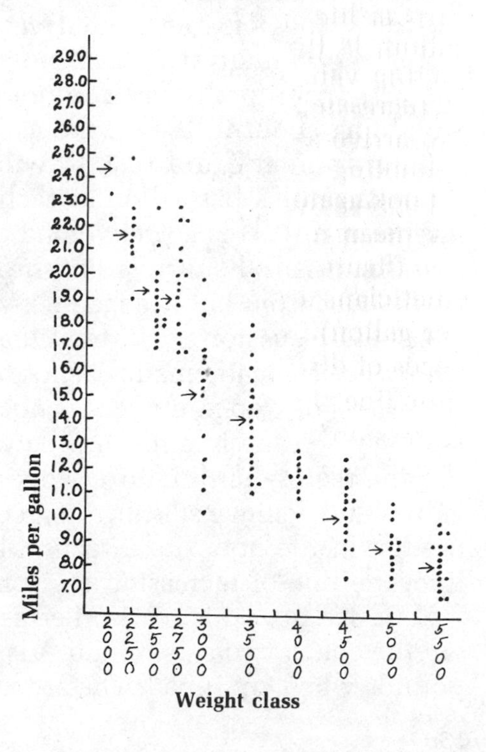

Fig. 10.3 Scatter diagram. This shows the distribution of miles per gallon for automobiles of various body weights.

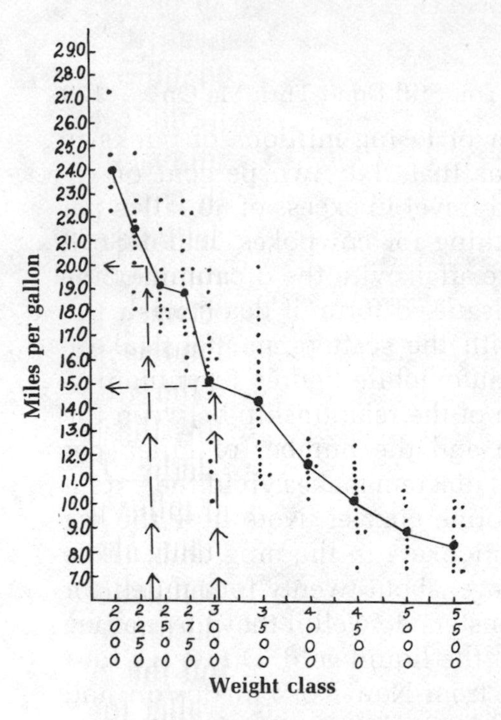

Fig. 10.4 Regression Line. An approximation of the regression line for predicting miles per gallon from the weight class of automobile (regression of *y* on x). The line is obtained by connecting the points showing the mean miles per gallon for each weight class.

The beauty of correlated data, particularly when the relationship is high, as in the case of weight of auto and miles per gallon, is that mathematicians have worked out ways of predicting values of a correlated variable. The method is known as *regression analysis*. Please don't let the term throw you. We can arrive at a pretty fair comprehension of regression without stumbling about in the arcane caverns of mathematics.

Look again at figure 10.3. Now let's draw a line that connects the mean miles per gallon at each weight of car. The resulting line (figure 10.4) is a pretty good *approximation* to what mathematicians call the regression line for predicting *y*-values (miles per gallon) from knowledge of x-values (auto weight). For purposes of discussion, we'll treat that line as if it were the regression line. In mathematical shorthand, it is called the line of regression of *y* on x. But that is not important. What is important is that when the relationship between x and *y* is high, we can use the regression line to predict *y* = values from known x = values and achieve startling degrees of accuracy. (We could also predict in the opposite direction by finding another regression line, the line of regression of x on *y*. But that's another story.)

Let's see how this works. Let us say that you are considering buying one of two cars. Brand A is flaming red and weighs 3,200 pounds when equipped. The second, brand B, is a metallic gold and weighs 2,400 pounds. You can make your best guess about the overall performance of brand A by looking at figure 10.3 and drawing a vertical line at 3,200 pounds until it intersects the regression line. Now look left to find the corresponding value for miles per gallon. Your best guess then, is that the 3,200-pound car will average somewhere around fifteen miles per gallon. Repeat the same procedure for brand B, drawing a vertical line at 2,400 pounds. That comes to about twenty miles per gallon, an improved fuel performance of about thirty-three percent, using fifteen miles per gallon as the base.

So there it is. Regression analysis can be relevant to your daily living. Indeed, although you may not be aware of it, you have probably provided a numerical value in more than one regression equation in your lifetime if you have ever applied for college admission, taken a series of tests for job placement, or filled out a life insurance application.

At the same time that regression analysis can be a tool of enormous value in the hands of a knowledgeable person, it can be an instrument of tyranny in the hands of an unthinking clod. This is what happens. The clod looks at the regression line and treats the predicted score as if it is God given and absolutely precise. He says such things as "Johnny Soinwhich got a score of five hundred sixty on the SATs. Since our cutoff score for admission is five hundred seventy, Johnny is obviously incapable of performing at the intellectual level demanded of our students. Now Mary Inlikeflynn is a different story. She got a score of five hundred eighty. Good girl. She'll do well at Dolc University." Our friend the clod forgets that all the data points fall directly on the regression line only when the correlation is perfect. In the real world, correlations are almost never perfect. This means that the data points are scattered about the regression line. (Take another peek at figure 10.2 to confirm what I am saying.) The lower the correlation, the greater the scattering, and the poorer the match between the prediction and the actual facts. Indeed, in any relationship that is less than perfect, about half of the actual scores will be lower than predicted and half will be higher. Thus, given the opportunity, Johnny Soinwhiches will occasionally perform admirably in spite of predictions to the contrary. And also on occasion, Mary Inlikeflynns will wind up on their tushes as they take the academic ten count. I remember well my freshman year in college. A delightful, bright, and attractive coed informed me, "I don't have to study. I have an IQ of one hundred thirty-five." Prior to the second semester of her freshman year, the dean generously offered to assist her in gaining admission elsewhere, after a suitable period of rustication.

All of this is another way of saying that there is a big difference between presenting a regression line (as in figure 10.4), complete with the dispersion of the original data points, and abstracting the regression line from the data (see figure 10.5).

The former permits the reader to observe the scattering of scores about the regression line, while the latter conveys the impression of great precision.

Correlation and Causation

But this is not the only way in which we can be deceived by correlational data. Consider the following facts and the conclusions drawn from them.

Fact 1: Individuals who lead active physical lives generally enjoy better health than those whose greatest exertion involves moving their duffs from one sedentary position to another.

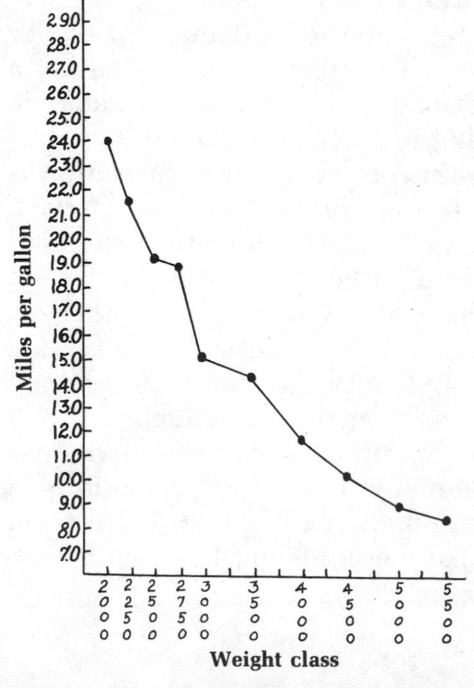

Fig. 10.5 Misleading Graph. This figure is the same as Fig. 10.4. However, the data points have been omitted. The result is the appearance of greater precision than the data warrant.

Conclusion 1: A regular routine of exercise is vital to good health. The more we exercise, within difficult-to-define limits, the greater the health we enjoy.

Fact 2: There is a positive relationship between the number of years that you go to school and the amount of money you are able to earn over a lifetime.

Conclusion 2: Get all the education you can. If you are a school dropout, you are also destined to become an economic dropout. On the other hand, if you get a college degree, your economic success is assured.

Fact 3: There is a positive relationship between the number of handguns produced each year and the number of murders committed by the use of this instrument.

Conclusion 3: We should reduce the annual production and importation of handguns since they are the cause of about two thirds of all murders. Since fewer murders by handguns are committed when fewer weapons are available, we can lower the murder rate by restricting the production of handguns.

All of these examples have something in common: They use the correlation between two variables as the basis for concluding that one causes the other. Let us examine each of them in more detail.

Let's look first at conclusion 1: A regular routine of exercise is vital to physical well-being. Being a tennis fiend, I am quite comfortable with this conclusion. It seems obvious that any well-planned and regularly practiced physical fitness program must be salubrious. This includes jogging, if we overlook the greater incidence of jogger's toe, jogger's heel, jogger's arch, jogger's ankle, jogger's knee, and jogger's emphysema, to name but a few. The folk wisdom "Sound body, sound mind" is also appealing. But what causes what? Are active people more healthy because they are more active? Or are healthy people more likely to lead active physical lives because their good health enables them to do so? These questions simply cannot be answered on the basis of correlations. Incidentally, the health and medical sciences abound in "causal facts" that are really statements of relationships among variables (see chapter 12).

Now let's take a look at conclusion 2. Figure 10.6 graphs the anticipated lifetime income for male wage earners twenty-five years old and over who have enjoyed varying numbers of years of education. The earning power of education appears most impressive. Men with four or more years of college can expect to earn, over a lifetime, about three times as much money as those who failed to complete elementary school. Surely there is no better evidence that education is a sound economic investment. Right? Nonsense. Don't get me wrong. I have nothing against education. It has kept many pairs of shoes on the feet of my children over the years. Nor will I argue with the view that education provides many nonmonetary values that one may cherish often between matriculation and Medicare. In fact, I shall gladly assert the latter view. I still ruminate over outstanding lectures on philosophy, ethics, and psychology that I heard over thirty years ago during my undergraduate years. From my view, the value of education need not and should not be measured by an economic yardstick.

But many people do. Their arguments go like this: "Look, if you don't get a college degree, you'll always be a schlep like Joe Snopes. You wanna be a schlep all your life? Not able to provide for your family? Welfare? Or do you want to be like Stan Upright. He's got a good college education and he's going places, let me tell you that."

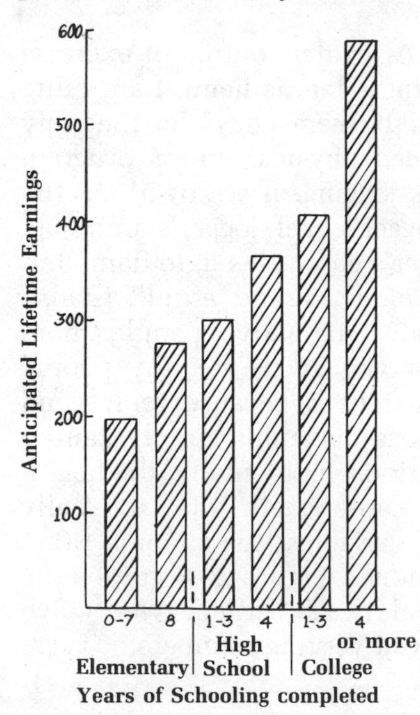

Fig. 10.6 Lifetime Earnings. Expected lifetime earnings of male wage earners 25 years of age and older. Over a lifetime, a college graduate can expect to earn about three times as much as a male who never completed elementary school. (Source: U.S. Bureau of the Census. Series P-60, No. 74)

The only trouble is that the data do not prove that education leads to or causes higher incomes. The data merely prove that educated people earn more. They may be economically successful for many reasons: They may be brighter or more highly motivated; they may be consummate con artists; they may have well-to-do parents or grandparents (a Rockefeller is likely to earn a healthy income during the course of a lifetime if he or she possesses no other skill than the ability to clip coupons from municipal bonds). Contrary to the view propagated in *The Wizard of Oz*, giving a college degree to a tin-head will not suddenly transform him into a computer capable of talking algebra. Stated succinctly, people who are educated differ in many ways from those who are not. Their economic success may stem from any one or a combination of all the ways in which they differ. It is no more appropriate to single out and credit education for the economic success of its graduates than it is to blame all of society's ills, as some do, on progressive trends in education or the "new math." (They learn that four plus two is equivalent to two plus four but do not know that either equals six.)

But that's not all. Take another look at figure 10.5. Do you think that each of the columns is representative of the same population of wage earners with respect to all important variables? In this day of compulsory education, what age group is most likely to have failed to finish elementary school? Youngsters, middle-agers, or those breathing hard on FICA? Clearly, the last of the three. But they earned most of these wages prior to double-digit inflation, at a time when, as the saying goes, a dollar was worth a dollar. I even remember when twenty-five dollars a week was considered a living wage. Such citizens could not be expected to earn a whole lot of money over a lifetime, no matter how much education they had.

Now let's look at the right-hand column of the graph. What proportion of our older citizens obtained college degrees? Very few, I assure you. To be more precise, about 3 out of every 10,000 U.S. citizens graduated from college during the year 1900. In contrast, about 5 in every 1,000 graduated during the year 1977. Relative to their number, about seventeen times as many great-grandchildren graduated from college as their great-grandparents.

So there you have it. Wage earners with low educational profiles are made up largely of older citizens who earned most of their income when wages were low. In contrast, those with high educational achievements are largely of the present generation, drawing income when wages are high. Their lifetime expectations are correspondingly high. So let's not justify education on pecuniary grounds. It simply doesn't wash.

In the example above, we are not justified in attributing a causal link between years of education and lifetime earnings because too many other variables are correlated with education—intelligence, motivation, and socioeconomic status of the family, to name a few.

In the language of statistics, these variables are *confounded*—they are so intertwined that there is no way of disentangling one from the other in order to ascertain their separate effects.

I am reminded of a story told to me by one of my friends in academia. It seems that one of his students had undertaken a term paper on the causes of alcoholism. After many weeks of intensive library study, the student uncovered a substantial difference among Catholics, Protestants, and Jews in their rates of alcohol addiction. It seems that Catholics are highest and Jews lowest. The student pondered these facts, asking such questions as, "In what ways do Jews differ from Catholics and Protestants?" Suddenly he was struck with the blinding light of cosmic insight: "Jews are circumcised; Catholics and Protestants are not. We can wipe out the scourge of alcoholism by circumcising everybody!"

To all of this, my professor friend raised but one question: "Do you think circumcision will be equally effective with female alcoholics?" In truth, Jews differ from Protestants and Catholics in many ways other than in the practice of the rites of circumcision—the ceremonial use of alcohol in early years, to name but one.

But what about the evidence, cited earlier, concerning the relationship between the annual production of handguns and the number of murders committed by gunfire? Surely, no one can question the validity of ascribing the increase in murder by guns to the greater availability of handguns. Or can one?

Let us examine the evidence. Table 10.1 shows the number and percentage of murders committed with various weapons

Table 10.1 Murder victims, by weapons used: 1963–1977 (Adapted from *The American Almanac: The Statistical Abstract of the U.S.* (New York: Grosset & Dunlap, 1976, 1980).

Year	Murder victims, total	% Guns	% Cuttings or stabbings	% Blunt objects	% Strangulations and beatings	% Drownings, arson, etc.	% All other
1963	7,549	56	23	6	9	3	3
1964	7,990	55	24	5	10	3	3
1965	8,773	57	23	6	10	2	1
1966	9,552	59	22	5	9	2	1
1967	11,114	63	20	5	9	2	1
1968	12,503	65	19	6	7	2	1
1969	13,575	65	19	5	8	2	1
1970	13,649	66	18	4	8	3	1
1971	16,183	66	19	4	8	2	1
1972	15,832	66	19	4	8	2	1
1973	17,123	66	17	5	8	1	2
1974	18,632	67	17	5	8	1	2
1975	18,642	65	17	5	9	1	3
1976	16,605	64	18	5	8	1	4
1977	18,033	62	19	5	8	1	4

between 1963 and 1977. There is no doubt about it. Aside from the automobile, the gun is our favorite weapon for abruptly terminating the lives of others. What is more, it is becoming increasingly popular with each passing year.

"But wait," you say. "Murder is on the increase, but so also is the population." Perhaps the two events are merely keeping pace with one another. So let's compare the rates of growth of both population and murder, using a technique we learned back in chapter 7. We'll calculate fixed-base index numbers between 1964 and 1977 using 1964 as our base year. The time indexes are shown in table 10.2.

When these index numbers are placed on a graph (figure 10.7), the picture is quite clear. The rate of increase in murder by gun is far outstripping the rate of increase in population. Although not shown here, it is also far outstripping the growth of population in that segment of the population that commits most of the murders (seventeen-year-olds and older).

So far we have established the fact that murder is on the rise and that guns are number one on the hit list. Now let's relate this increase to the number of firearms produced and imported each year. Presumably, greater production means greater availability and greater availability means greater opportunity and

Table 10.2 Fixed-base time ratio index numbers between 1964 and 1977, using 1964 as base year (several sources)

Year	Murder by gun	Fixed base 1964	Population (in millions)	Time ratio index 1964
1964	4,393	100	191.9	100
1965	5,015	114	194.3	101
1966	5,660	129	196.6	102
1967	6,998	159	198.7	104
1968	8,105	184	200.7	105
1969	8,876	202	202.7	106
1970	9,039	206	204.9	107
1971	10,712	244	207.1	108
1972	10,379	236	208.8	109
1973	11,249	256	210.4	110
1974	12,474	284	211.5	110
1975	12,061	275	213.6	111
1976	10,592	241	215.1	112
1977	11,274	257	216.5	113

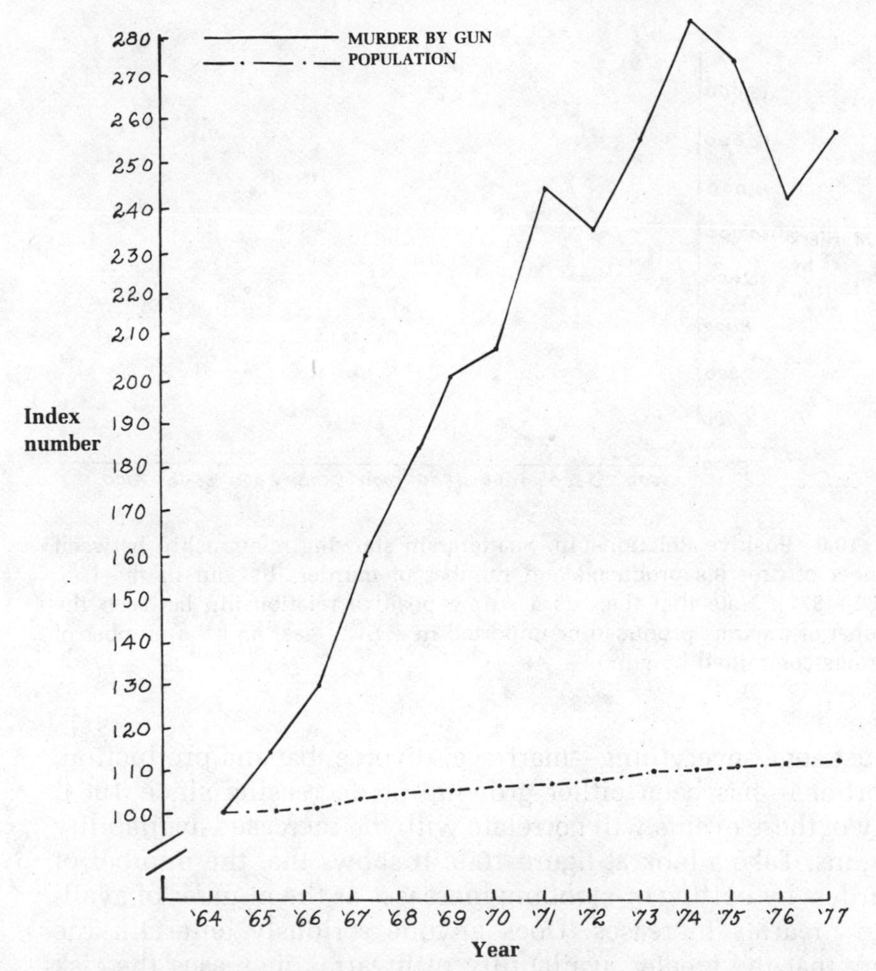

Fig. 10.7 Outstripping Graph. Graph of the time ratio index numbers showing increases in murder by gun and increases in population. Note that the increase in murder by gun is far outstripping growth in population.

temptation. The graph is shown in figure 10.8. Pretty convincing evidence, huh?

'Fraid not. Please don't get me wrong. I am not trying to add ammunition to the arsenal of the have-guns-will-travel crowd. Quite frankly, I think our insistence on the right to bear arms is a racial memory from the Neanderthal days. But my own feelings are irrelevant to the issue. If the weight of evidence is the high correlation between availability of guns and murder by gun, I must admit to being a skeptic. Why?

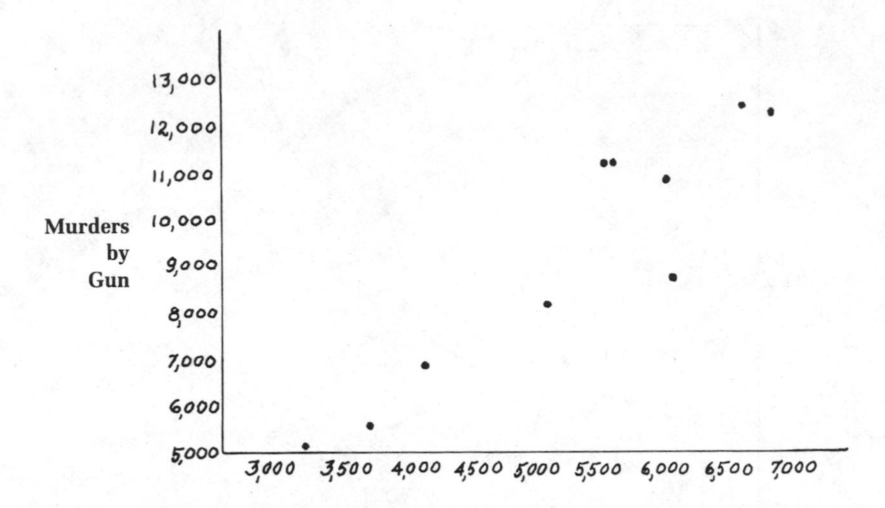

Fig. 10.8 Positive Relationship. Scattergram showing relationship between number of firearms produced and number of murders by gun in the U.S. (1965–1977). Note that there is a strong positive relationship between the number of firearms produced or imported in a given year and the number of murders committed by gun.

Just about everything—marriage, divorce, banana production, abortions—has been either growing or decreasing since 1964. Any of these events will correlate with the increased availability of guns. Take a look at figure 10.9. It shows that the number of murders by *cutting* or *stabbing* increases as the number of available *firearms* increases. Does anyone seriously entertain the view that the greater availability of firearms increases the risk of death by stabbing, or that we can reduce the number of gunshot murders by cutting down on the availability of knives? And how about figure 10.10? We see a beautiful negative relationship between the number of farm workers and the number of murders committed by guns. If we take this sort of evidence seriously, we had better figure out how we can get them back on the farm, even after they have seen Paree.

Errors of this sort are far more common than we would like to believe. Consider the following cases.

1. An aspirant for political office proclaims, "There has been a twelve-percent increase in crime since my opponent, the incumbent, came to office." To ascribe a causal link between the

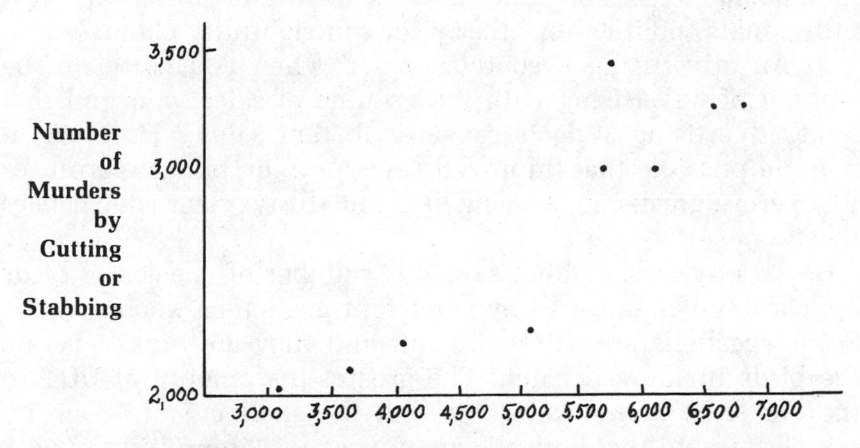

Fig. 10.9 Another Positive Relationship. Scattergram showing relationship between number of firearms produced and the number of murders by cutting or stabbing. Note that there is a strong positive relationship between these two variables. However, it would be difficult to build a case for the view that one causes the other.

Fig. 10.10 Negative Relationship. Scatter diagram of relationships between number of farm workers and murder by gun. The diagram depicts a strong negative relationship: As the number of farm workers decreases, the number of murders by gun increases.

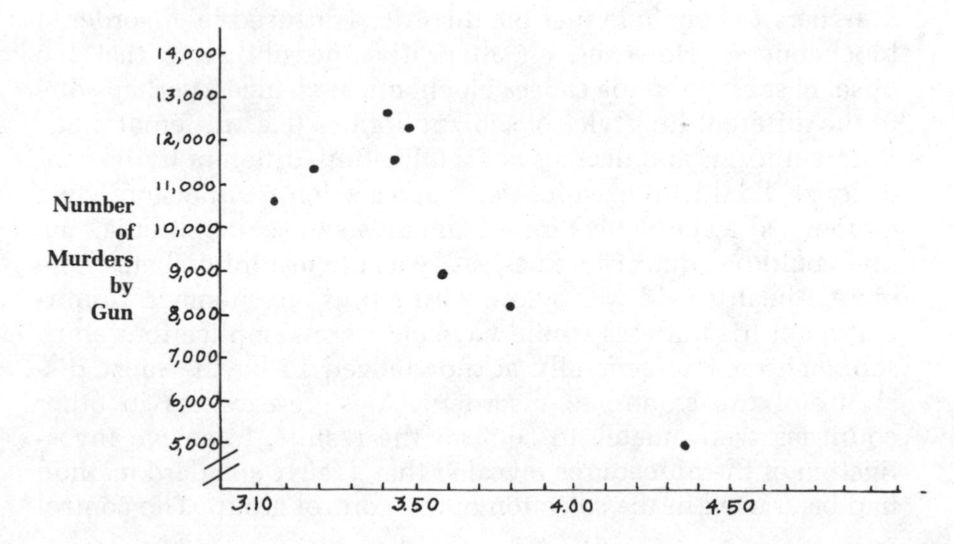

incumbent and crime is to assert a degree of power that few individuals, political or otherwise, can rightfully claim.

2. An advertising executive says, "When you correlate the amount of advertising with the volume of sales, you find that large advertising budgets cause exuberant sales." However, it is quite possible that improved sales provide budget surpluses that permit greater advertising. It is not always clear what causes what.

3. An environmentalist says, "A number of species of birds appear to be endangered by persistent pesticides, such as DDT. Some eggshells are extremely soft and they are broken before the birds inside can hatch. The greater the amount of DDT in the yolk, the more likely it is that the shell is soft." One researcher found that both the amount of DDT found in the yolk and the softness of the shell correlated with the nearness to civilization. He proposed an alternative explanation for the soft shell phenomenon, namely, the birds close to civilization were more likely to be frightened by bird watchers (few of whom venture into remote locales) causing them (the birds, not the watchers) to lay their eggs prematurely.

4. There is a crippling emotional disorder known as schizophrenia. When blood samples from schizophrenic patients are compared with samples taken from normal people, numerous biochemical differences are observed. This relationship between biochemical events and schizophrenia has prompted some researchers to conclude that the disorder is caused by disordered biochemistry. However, an alternative possibility is that the onset of schizophrenia causes biochemical changes, perhaps due to the different life styles of schizophrenics (e.g., different diets, different forms and degrees of socialization, different living conditions). To illustrate, some years ago a scientist in Scandinavia announced a simple test for the diagnosis of schizophrenia, one that could be administered as easily as the test for diabetes. The mental health field was aglow with suffuse excitement. Confirmation of his findings would have enormous implications since schizophrenia is generally acknowledged to be the most disabling of the emotional disorders. Alas, researchers in other countries were unable to confirm the results. Intensive investigation of the procedures revealed that a fairly standard method had been used in the selection of the control group. The control

subjects, drawn from the "normal" population, were pretty well matched with the patients on such variables as age and sex. However, the patients in the hospital perforce took their meals in the institution; control subjects ate their meals at home. It turned out that the diet of the patients was deficient in citrus fruits. Suddenly, it was realized that the scientist had, in fact, developed a diagnostic test. However, instead of schizophrenia, it detected vitamin C deficiency!

5. New York State recently banned total nudity, including bottomless entertainment, in establishments purveying alcoholic drinks. The reason? According to liquor authority spokesman Michael Roth, because of "years of experience that have shown nude dancing and similar entertainment frequently lead to prostitution and other kinds of sexual conduct between performers and customers."[1] As an alternative hypothesis, what about the possibility that women willing to bare their bottoms in public might also be willing to share their bodies in private? It is also possible that the clientele of these establishments have other-than-visual entertainment in mind, and are somewhat more aggressive in pressing their point.

There is a lesson in all of these examples. Causation is not easy to establish. Correlation is a necessary but not sufficient condition for establishing a causal relationship. If events A and B are not correlated, it is hardly possible that one causes the other. The experiment is widely hailed as the one unambiguous means of establishing a causal link. In an experiment, two or more similar groups are treated alike in all ways except with respect to the administration of the experimental condition. If differences in behavior emerge, it is presumed that the experimental treatment caused the differences. To qualify as an experiment, one vital condition must bc achieved: It must be possible to manipulate the experimental conditions so that they can be administered to subjects who are selected at random from some population and assigned at random to experimental conditions.

Unfortunately, many so-called experiments fail to qualify with this absolutely essential condition. The reason is that the so-called experimental treatment is often not under the control of

[1] The Associated Press release, December 5, 1975.

the experimenter. It is not free to vary. In a sense, the subjects select themselves for the so-called experimental treatments. For example, in studies of intelligence of whites and blacks, it is not possible for the experimenter to say, "For the purposes of the study, I will make you a black and you a white. We will then compare your intelligence scores." In the real world, whites and blacks already exist as different individuals. They differ in so many ways that it is not possible to say whether or not differences in scores on IQ tests reflect differences in underlying intelligence or differences in prenatal care, diet, cultural values, early experiences, motivations, language, or what have you.

Whenever you see the results of a so-called experiment, get in the habit of raising the questions, "Is it an experiment in the strictest sense of the word in which the experimenter exercises control over the administration of experimental treatments? Or is it, rather, a correlational study that masks under the guise of an experiment?" If you do, you'll find that an astonishingly large quantity of "conclusive results" are still very much in the speculative stage of their development.

CHAPTER **11**

Giving Business the Business: Computer Crime

Another day, another pressure cooker. Construction of the main computer facilities for the First Real Bank was nearing completion, and there was still so much to be done. Sly Sithin, the programmer for the facility, was beginning to show signs of raggedness around the edges as he labored day after dreary day at debugging the system. To add fuel to the fire, the board of directors of the bank had announced a grand opening celebration to take place in precisely two weeks at 10 A.M. sharp. Everything had to be ready for opening day, and that meant everything. There could be no mistakes, no careless errors with decimal places. Sly would lie awake at night bathed in a pool of sweat as wild images of computer failures raced through his mind. There was that case in which a computer issued monthly checks to a life insurance beneficiary that ran intc the hundreds of thousands of dollars. Or was it millions? The face value of the insurance policy was just about enough to provide a good funeral for the deceased. Frightening. Absolutely terrifying. He must be absolutely certain nothing like that could happen with his baby. He just couldn't be careful enough. Today, like yesterday and many yesterdays before, would be spent in simulating various types of computer crises. If the computer behaved badly, he would have to rethink the whole subroutine, find out where it went wrong, and make necessary adjustments. If only he had more time . . .

"How's it going, Sly?" It was the bank president.

"Fine, Mr. Martinez. A little hectic, but fine."

"Good. You will have it ready by opening day!"

"I sure hope so."

"You misunderstand. I was not asking a question. I was telling you. You will have it ready by opening day."

"Yes, I sure will."

"Good." He hated to sound so hard with employees, but he couldn't afford to relax his vigilance a single moment. This computer had made him particularly nervous. In the past, he always felt in control of the operations of the bank. Starting, as he did, at the bottom of the rung, he had acquired a fine working knowledge of every department. But the computer was a different breed of animal. Even had its own language. All he could do was hire the best and trust to God that they knew what they were doing.

"Mr. Martinez, a quick question before you leave. I'm working on the interest subroutine now. How do you want me to round?"

"Round?"

"Yes. You know the interest is never correct to the penny. There is almost always a remainder. Should I round to the nearest penny?"

"Do whatever is easiest. No, on second thought, round down. We should hold on to those fractions of a cent. After all, they do add up." A laugh and departure.

The exchange began turning some wheels in Sly's mind. "On the average," he thought, "the remainder will be about a half a penny per account. That's not much. Why should the bank get it? It doesn't belong to them." Bit by bit, an audacious plan began to evolve. "There are over two hundred thousand different savings accounts, that would amount to . . . let's see . . . over one thousand dollars a month. Whooee." He felt his heart beat just a little faster and perspiration began to form in the palms of his hands. "Now if I modify the program ever so slightly . . . yes, I'll round to the lowest cent, assign the remainder to my account . . . a thousand extra each month. No one will be hurt, and it will never be noticed."

Indeed, it was noticed, but it took several years to apprehend the programmer. This fictionalized account described one of the earliest computer crimes.[1] Moral considerations aside, it has all the elements of a favorable risk-taking situation (See chapter 1). The long-term gain can be phenomenal (in the millions of dol-

[1] Davis, W. S., *Information Processing Systems* (Reading, Mass: Addison-Wesley Pub. Co., 1978) p. 393.

lars), the risk of losing is extremely low (perhaps fewer than one in a hundred are caught and one in a thousand receive a jail sentence), and the turnover rate is so large that infinitesimal percentages can be turned into enormous gains. It is la crème de la crème of white-collar crime since intimate knowledge of computer technology is essential for its successful accomplishment. It is well beyond the capabilities of most nine-to-five brown paper baggers, requiring intelligence, imagination, and a peculiar brand of chutzpah. Computer thieves are not dangerous people in and of themselves. They inflict no physical harm on their victims, do not terrorize them, and use only their specialized knowledge as weapons. Yet, in a fraction of a second, they can commit crimes that put the Brink's robbery to shame. Why, in 1979, the average loss involving financial shenanigans with a computer was a whopping $450,000 and going up![2] Moreover, the victims may never know they've been had.

Statistical inferences concerning the incidence of computer crime are little more than informed guesses. The reason is simple. Many computer crimes go undetected because the crime is as sophisticated as the equipment. The perpetrators use all avenues of vulnerability—from data entry to internal circuitry of the computer, to the programs that give commands to the computer.

Those that are discovered are usually either poorly conceived and executed or somebody stumbles on them by accident. For example, an insurance company increased its profits by selling false policies to other insurers. In this case, the computer conceived and gave birth to the fraudulent policies. These were added to the legitimate policies of the company. When audited by a firm with a national reputation, the records of the company were certified to be accurate. It was only when a disgruntled employee ratted on the company that the fraud was discovered.[3]

In the few decades since the inception of the computer, many special computer languages have been devised—COBOL, BASIC, FORTRAN. Not to be outdone, a colorful vocabulary has evolved to describe different types of computer crimes: *data diddling, the Trojan horse, the salami technique,* and *superzapping,* to name but a few. Let's look at some of them.

[2] Henkel. T., "DP Crime Seen Hitting New Heights," *Computerworld,* V. 13, 45, November 1979.
[3] Davis, *op. cit.*

Data Diddling

When you diddle with data, you deliberately doctor it prior to delivering it to the computer. It is simple, safe, and the most common method used in computer crime. "The changing can be done by anybody associated with or having access to the processes of creating, recording, transferring, encoding, examining, checking, converting, or transferring data that ultimately enters a computer."[4]

Let's look at a typical example. A timekeeping clerk working in a large department of a railroad company was responsible for filling out data sheets showing employees' weekly hours worked. The information fed into the computer included both the employee's name and identification number. The clerk made an interesting observation: The computer recognized only the identification number and used this number to "look up" the employee's name and issue a pay-roll check. Fellow employees, however, paid no attention to other employees' numbers. They recognized and referred to each other by name only. Moreover, all hand processing was done by name rather than by number.

"Yes, that's the job I'd like to apply for—the Time Keeping Clerk spot that paid $33,000 last year."

The clerk devised this simple scheme. He identified a number of employees who frequently worked overtime. He filled out an overtime form for several employees each week (those who did not work overtime during the week) but assigned his own number. People processing the forms each week might think, "Oh, John Doe worked overtime again." But not the computer. It ignored the name, read the employee number, and issued a check to the jolly timekeeping clerk.

How was the thievery discovered? It would be nice to say that the computer program was sufficiently sophisticated to pick up the diddling—or auditors of the company did some excellent detective work—but both of these would be untrue. After several years in which the clerk successfully embezzled thousands of dollars from his employer, his W-2 form was, by chance, looked at by an IRS auditor. "An awful lot of money for a clerk to be making," he thought. He called the clerk's employer, perhaps to apply for a position as assistant clerk. The employer agreed that something was rotten in the railroad. When confronted with the evidence, the timekeeper broke down and admitted his guilt. This is a case of poetic justice: The villain catches the villain.

The Trojan Horse and Logic Bombs

Remember the original Trojan horse? After failing to subdue the valiant soldiers of Troy by direct frontal attack, the Greeks constructed a giant hollow horse and filled it with armed soldiers. When the Trojans welcomed it inside their gates, the horse disgorged its contents. The sack of Troy ensued. Beware of Greeks bearing gifts.

Also beware of computer programs that contain wide-open spaces where additional instructions can be added with nobody the wiser. A business program, for example, may consist of over 100,000 different instructions and data. The operating system may itself contain millions of different instructions. A knowledgeable computer expert can insert some instructions to execute commands whenever the target program comes on (e.g., the business program). In essence, the hidden instructions act like a highly trained assassin. It lies quietly in wait for the target to

[4] *Ibid.*

appear, executes the instructions in a few thousandths of a second, and then removes all evidence of its existence. The instructions may be to transfer funds or equipment, to print information desired by a competitor, or to sabotage the system. Go find the criminal! One payroll-system programmer used the Trojan horse method to build a one-time-only secret code into the system. This code might best be labeled REVENGE. This is how the coded instructions read. If his name was ever removed from the personnel file (meaning his employment had been terminated), the secret code would cause the entire personnel file to be erased. Any questions why this technique is referred to as the logic bomb?[5]

I have even heard allegations that some technicians build malfunctions into their systems. As an act of sabotage? No, to assure long-term employment. This is how it goes.

> CUSTOMER: The cathode ray tube just went out on me.
> MANAGER: Frank, call Carl. He's a genius at trouble-shooting this equipment.
> CARL: You rang?
> FRANK: Get over here quick. We've got a malfunction.

Carl appears on the scene a few moments later. He appears to be deep in thought. As a matter of actual fact, he is. He plans to use the computer to simulate a gambling situation involving the weekend football games. He'll run the program, ostensibly "to check out the computer system." No one will guess what he actually is doing.

> CARL: What's the trouble? Cathode ray tube? Hmm, that's a new one.

Actually, it's not a new one. It's one of several malfunctions he has built into the system, a tiny logic bomb that explodes whenever a specific target appears on a program. Of course, he knows immediately what is wrong and is able to correct it in minutes. But before doing so, he goes through an elaborate pantomime of checking switches, meters, and dials. Then he announces his success.

[5] *Ibid.*

CARL: That should do it.

Immediately, the cathode ray tube lights up.

FRANK: That Carl's a genius.

CUSTOMER: I can't believe it. He went right to the heart of the trouble. I'll bet you pay dearly for his services.

MANAGER: You'd better believe it, and he's worth every last cent. Around here, we call him Merlin.

CARL: It'll be another five minutes. While I'm here, I'll run this check program to see that everything else is functioning OK.

MANAGER: Good idea.

The manager adds, aside to the customer, "He's so thorough. Always runs a check program after a malfunction."

In another case, secret instructions were inserted into the operating system of the computer. A routine was activated periodically to check the year, date, and time of day. Precisely two years later, at 3 P.M., the time logic bomb "exploded." It confessed a computer crime on all 300 computer terminals on-line at the time and then caused the system to crash.[6] By that time, of course, the criminal was long gone.

The Salami Technique

Remember the round-down fraud accomplished by Sly Sithin? Well, he used the *salami technique*. The name derives from the fact that very small slices of the total salami are taken. In other words, you slice it thin. It is highly unlikely that auditors will notice the peccadillo unless they follow through every computer program step by step. An unlikely event, since this is a monumental task in large facilities.

A common variation of the salami technique uses the Trojan horse to gain access to a loosely constructed program and to insert a minor change in instructions. For example, you may insert a subroutine that instructs the computer to randomly select a few hundred out of thousands of accounts, randomly

[6] *Ibid.*

subtract small amounts (averaging, say, ten cents) from each, and then transfer the money to an account from which you may make a legitimate withdrawal. Since no money is removed from the system of accounts, no red flags are raised by the computer. From the point of view of the computer, the whole transaction is on the up and up. After all, like a good soldier, it just follows orders. Now, imagine you are the victim of such a crime. What do you do? Rush down to your local bank, fire in the eyes, and demand that your account be credited with the missing dime? Of course not. What with deposits and withdrawals occurring throughout the month and fees charged on your average balance or when the balance drops below a certain dollar value, you are probably quite elated if your calculations come within a dime of the bank's statement. To be utterly truthful, I feel like Superbrain if the disparity is less than a few dollars.

Even if you were to complain, you might get an answer like the following.

"You know, it's all done by computer. The computer doesn't make mistakes."

"You're saying *I* made a mistake?"

"No, I didn't say that. But if you go over your additions and subtractions, you may find that an eight became a nine, or a five a six. Something like that. It's easy to do."

"I have all my checks with me and your statements of charges—the whole kit and caboodle. I'd like to go over everything with you."

"Everything?"

"Yes, everything."

"Well, if you insist."

The two of you laboriously pore over every last statement in the ledger. An hour later, you emerge triumphantly.

"I told you there was an error. Computers don't make mistakes! Pha!"

"Yes, there was a disparity of ten cents, I must admit. Probably an error made by a card punch operator."

"So what are you going to do about it?"

"We'll credit your account with the ten cents."

"What else?"

"Like an apology?"

"No, like an investigation."

"An investigation for a ten-cent error? That's hardly likely.

Now, to get your credit, I want you to fill out form 10.43G in triplicate, form 9056 in duplicate, and form 0913, section 8 in quadruplicate . . ."

Superzapping

Imagine the following scenario. You find yourself locked in a security-tight building—steel doors equipped with time locks, and all windows barred. In a word, no way in or out. You suddenly become aware of smoke and heat emanating from one of the rooms in the building. What do you do? You panic, of course. When that doesn't help, you search desperately for that little red box that reads "Break glass in case of emergency."

Some computers are like that security-tight building. When an emergency (i.e., malfunction) occurs that cannot be rectified by standard recovery or restart procedures, it is necessary to use a "master key" that provides access to the normally secure contents of the computer. In many IBM facilities, the program that "opens secure doors" is called Superzap. The problem is that once the doors are open, all controls are suspended. Nothing is wrong with that as long as the person doing the superzapping is dependable. Otherwise, it's like opening the bank vaults to

"You stand accused of perpetrating the Trojan Horse, salami technique, logic bombs, and superzapping on the computer. What is your defense?"

burglars and giving them free rein. However, with a computer, a great deal of mischief can be accomplished in fractions of a second. In one case of superzapping, the manager of computer operations in a bank was directed by management to correct errors that had crept into his system as a result of sloppiness during a computer changeover. He soon discovered that in the absence of the usual controls, he had free rein to falsify records and transfer funds, without leaving evidence of changes in the data files. He switched over $100,000 to the accounts of three of his friends before a customer's complaint alerted authorities to the fact that all was not well in computerland. The four were subsequently indicted and convicted.[7]

Scavenging

With the computer, as with so much else in today's world, time is money. In many computer systems, magnetic discs or magnetic tapes are used to store information temporarily while other operations are proceeding. Since it requires valuable computer time to erase the data held in temporary storage, many systems permit the information to remain until new data are written over it. As you might imagine, there are some enterprising individuals who will take advantage of this Achilles' heel in the system to acquire valuable information free of charge.

A time-sharing service in Texas boasted a number of oil companies as customers. Among other things, they used the computer to analyze the seismic data obtained from various potential drilling sites. This is valuable information, so closely guarded that even investigative branches of the United States government are denied access to it. One of the regular customers of the computer service found a way to accomplish what others could not. On arrival, his first request was always that a temporary storage tape be mounted on the tape drive. The computer operator noticed that the customer activated the "read-tape" light before he had written anything on the temporary storage tape. That was very strange. Strange indeed. "What would he be reading if he hadn't as yet written anything?" he wondered. Finally, after a number of incidents of this sort, the operator called this

[7] *Ibid.*

unusual behavior to the attention of management. It did not take long to discover that the customer was engaging in industrial espionage. He was reading the temporary storage tapes of various oil companies, printing out their data, and selling the information to competing oil companies. Is there no honor among thieves?

A Few Statistics about Computer Crime

These are but a few of the methods by which computer crimes are initiated. Others include *trap doors, asynchronous attacks, data leakage, piggybacking,* and *impersonation*.[8] All result in one of four classes of abuse: physical destruction, intellectual property deception and taking, financial deception and taking, and unauthorized use of services.

How rampant is computer crime in this country? Who knows? It's an entirely different breed of cat. It is committed by extremely intelligent individuals who frequently take more pleasure in the fact of the crime than in its proceeds. They find the challenge exhilarating. As we have seen, the crimes may take place in mere fractions of a second, involving either millions of dollars or extremely sensitive and valuable information, and then be erased, leaving no permanent record of the larcenous transactions. Who's to catch the perpetrators? The people in higher management of large corporations, banks, insurance companies, to name a few, couldn't care less about *how* a computer functions. For them, a computer is merely a high-speed business tool. They may be mystified and totally frustrated when unexpected losses occur, but be completely at sea for an explanation. Most auditors are of little use since in a cleverly executed crime the books will balance. Beyond that, very few prosecutors and police can boast of being computer experts. They fear that the defense attorneys will make monkeys out of them during a court trial.

So what happens? The best we can guess is that many computer crimes go undetected. Some experts have estimated that

[8] For a discussion of these techniques and the methods by which they may be detected, see *Computer Crime: Criminal Justice Resource Manual, SRI International, 1979*. Copies are on sale by the Superintendent of Documents, U.S. Government Printing Office, Washington, D.C. 20402. The stock number is 027-000-00870-4.

less than one in a hundred is detected. But if I were to say, "No, it's less than one in a thousand," it is unlikely that anyone would seriously challenge me. Statistics on computer fraud have many of the characteristics of the disembodied statistic (chapter 5). They sound very precise (99.44 percent precise, to be precise). Particularly alarming is the fact that accidents, rather than technical investigative activities, have uncovered many of the crimes. With increasing proficiency in plying their trade, many computer thieves may be able to reduce the incidence of accidental discovery. Perhaps they already have and what we have identified as computer crime is just the tip of the iceberg (see figure 11.1). What then?

Even if detected, many will go unprosecuted. Few DAs are gung ho to enter the word jungles of *COBOL, FORTRAN, compiler, flip-flop circuits, CPU, disc storage, DBMS, I/O Bound,* and *JCL.* And how many corporations are willing to admit that they have been successfully swindled? Indeed, they may just turn around and offer the culprit a fat contract as a consultant on computer security. The guiding philosophy appears to be: It takes a thief to catch a thief or If you can't beat 'em, join 'em.

Statistically speaking, computer crime is, at this moment, a

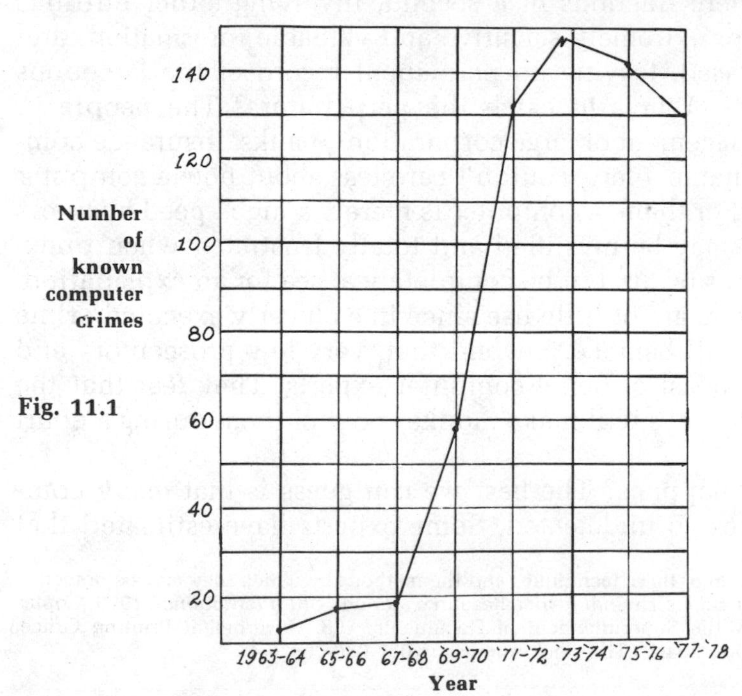

Fig. 11.1

dark cave with no light apparent at the end of the tunnel. Public indifference probably stems from the fact that computer criminals are romanticized as lone wolves bucking the system. Who has not, at one time or another, wanted to put a bank president in his place or get even with a computer that long after we have sent a check continues its nasty letters saying we are behind in our payments?

"After all," you say, "it doesn't impact on me."

"Oh, yeah? What if, at this very moment, a large corporation in which you have nested your retirement egg is slowly being drained dry by a computer thief? What about that over-the-counter stock that your broker has difficulty tracking down? Does it exist only in the mind of some computer somewhere? What about diddled data that certify the braking system on your latest dream machine is 'safe'?"

"What you are saying is that computer crime is not all fun and games?"

"I am saying more, so much more. Computer criminals can kill."

"What, with logic bombs? Come now, you can't gain my support by overstating the case."

"I don't think I am overstating the potential dangers. Think of what we have seen in these few pages. Computer crimes can take place in a few thousandths of a second. The criminals can erase all evidence that a crime has taken place; corporations do not like to prosecute computer criminals, even when caught red-handed; the police and prosecutors are afraid to touch them. In short, as observed earlier, computer crime is a magnificent risk-taking opportunity. Its payoff is big. The risk is low. You may even get a high-paying job if you are caught."

"You're beating around the bush; haven't answered my question. How can computer thieves kill?"

"The potential is there. Think for a moment of the sensitive information about each of us that is presently stored in discs or tape at the Department of Justice, the Internal Revenue Service, the Census Bureau, various credit agencies, and a half a hundred other agencies and organizations. Much of this stored information is unevaluated. Somebody may have said bad things about you or me that have no substance in truth. But they are there. Or perhaps you once committed an indiscretion for which you have already paid your share of grief and guilt. Are you willing to bet that the information is not sitting on a shelf somewhere?

"Have you thought of the potential for extortion if someone gained access to the information? If a person suddenly sees his world crumbling about him and takes his own life, who is guilty of murder?"

"I get your point. For that matter, computer thieves have the potential of destroying businesses. Think of all the shattered lives that would involve."

Shattered lives, shattered businesses, and lowered productivity. How great is the risk? No one can answer that question, since no one really knows to what extent computer crime is occurring. It's a sobering thought that a computer sitting somewhere in this land may, at this very moment, be quietly and unobtrusively obeying commands that will transform your life in ways that, given the opportunity, you would summarily reject.

CHAPTER **12**

The Joy of Dying: Statistics and Health

Trim That Fat

For years we have been warned of the dangers of obesity. we have been told that it increases the risk of diabetes, high blood pressure, elevated levels of blood fats and cholesterol, strokes, heart attacks, and for all I know, fallen arches. Life insurance companies regard our corpulent brothers and sisters as so flawed that they are placed in a special-risk category. This means that they pay higher premiums for a given amount of coverage.

Of course, this is eminently reasonable. Everybody knows that we should trim the paunch, reduce cholesterol in our diets, and consume fewer animal fats. If there is any health fact that is established beyond any reasonable doubt, it is this: Fat kills. Oh, yeah?

"Not so," says Dr. Andres of the National Institutes of Health.

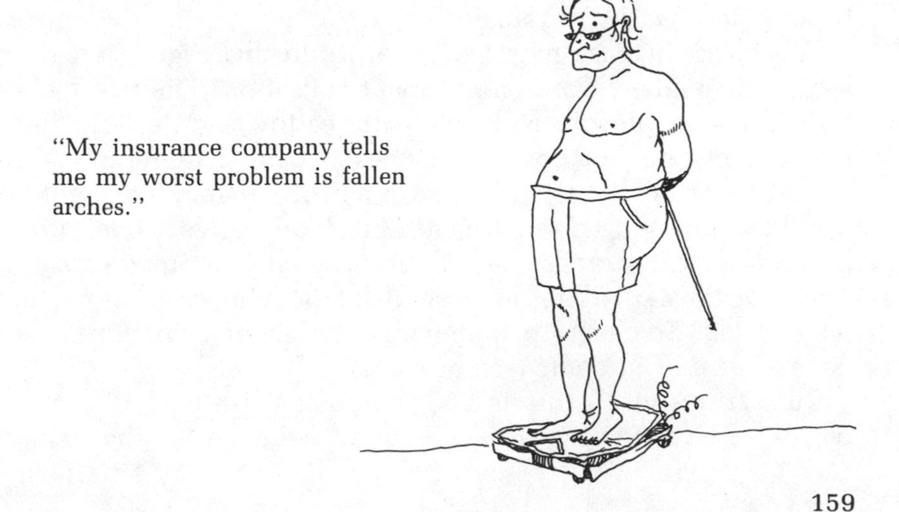

"My insurance company tells me my worst problem is fallen arches."

Following an intensive review of seventeen long-term health studies, he was unable to find a single instance in which the death rates were higher among the obese. Just the opposite. He reported that in all age categories, moderately *underweight* individuals had a *higher mortality rate* than their counterparts who were moderately overweight (including people 30 percent over ideal weight). How could this be? How could years of entrenched scientific belief be so far off the mark?

The answers to these questions are at once simple and complex. There was obviously something inadequate in the original study, performed in the 1950s, that gave obesity such a bad name. But what?

In a word: biased sampling. We have discussed this problem before. Recall the *Literary Digest* poll that gave Alf Landon a landslide victory in 1936? The sample did not represent the general population, but reflected only a very select segment of the population—the affluent. We know that in order to draw a conclusion about a population from a sample, the sample must be representative of that population. How does all of this apply to the landmark study that until this day still punishes the fat guys and gals for not exercising greater restraint at the dinner table? It seems that the sample in the study was composed of individuals who had been denied insurance because they were *already* health risks. Many of them died at early ages, as health risks are inclined to do. Many were also overweight. This is not at all surprising. People looking at the grave probably experience many changes in life style, including, perhaps, a tendency to become less active physically.

Whatever the case may be, we are into that regression thing again. Remember? Correlation is not causation. The two facts—unhealthy people die early, and unhealthy people tend to be overweight—do not prove that obesity causes their unhealthy condition. And it certainly doesn't prove that overweight individuals enjoying normal health are likely to keel over at any minute. Isn't it strange that in the original insurance study no one thought of looking at normal fat people to see how they were faring? So cheer up, members of Fat Liberation, you'll soon have your day in court (see box 12.1).

But seriously, all this is very confusing, isn't it? Whom are we to believe, in this day of experts, when even the experts

Box 12.1 Fat Not Necessarily Fatal; Skinny Is Scary

"Dick, excuse the interruption."

"Yes, Hal."

"Here's one hot off the presses. In a recent issue of the *Journal of the American Medical Association,* mortality rates are reported for people ranging from lean to obese."*

"And how do our porcine friends make out in this study?"

"Very well."

"Really?"

"On second thought, delete the very."

"Oh?"

"The sample in the study was divided into five build groups by both height and weight. Those in group one were extremely light for their height, while those at the other extreme (group five) were butterballs."

"And you say the butterballs made out well?"

"With reservations. They evidenced elevated blood pressures, higher serum cholesterol levels, greater amounts of uric acid with its associated gout, and higher rates of diabetes. But—"

"Well, chalk one up for the lard-asses. You've already deleted the *very* from *very well.* Now let me delete the *well.*"

"You didn't let me finish."

"Didn't you say enough?"

"No. In spite of all these conditions associated with obesity, the leanest people enjoyed the highest mortality rates."

"Now wait a minute. Am I hearing you correctly?"

"The skinny people enjoyed the highest mortality rates."

"Well, I hope they didn't enjoy it too much."

"Actually, it's all rather mystifying. Look at this figure. As you can see, the age-adjusted mortality rate is highest for the men in body group one. These are the string beans."

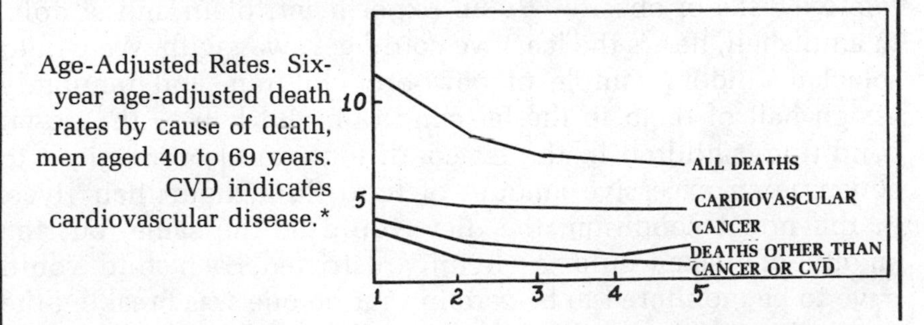

Age-Adjusted Rates. Six-year age-adjusted death rates by cause of death, men aged 40 to 69 years. CVD indicates cardiovascular disease.*

* Sorlie, P., Gordon, J., & Kennell, W. B. *Body Build and Mortality: The Framingham Study,* Journal of the American Medical Assn., Vol. 243, No. 18, May 9, 1980.

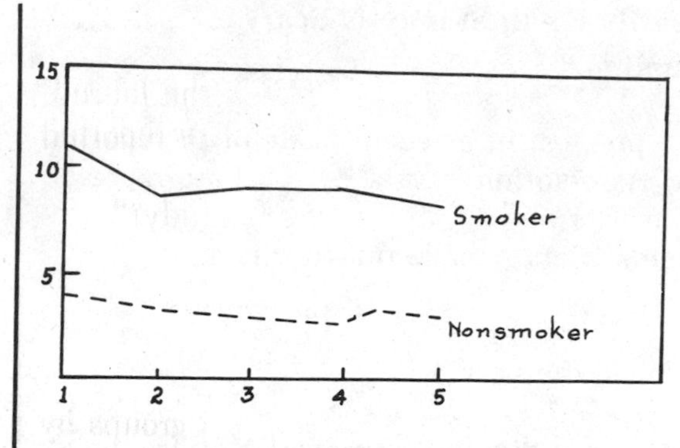

Smoking Status. Six-year age-adjusted death rates by body build and smoking status.*

"Amazing! And there is no explanation?"

"For a while they thought it was due to smoking. About eighty percent of group one smoke, whereas only fifty percent of obese men (group five) are addicted to the weed. Now, if you plot mortality rates by smoking status, you find that smokers generally have higher mortality rates. But, as you can see, men with that lean-and-hungry look fare poorer whether or not they smoke."

"I've got it! I know why the lean men did worse."

"Great! Let's hear it."

"Many of the men in group one were lean because they were already suffering from a terminal illness. Except for the fact of their poor physical health, they would not have been in group one."

"Good thinking. However, this idea was checked out by the researchers and was found to be largely without merit."

"So, where does this leave us?"

"Rejoicing over the fact that I don't have to go on a diet."

cannot agree? How could we establish for once and for all the relative risks of obesity? By an experiment, plain and simple. In a nutshell, here's the least we could get away with. We would select a random sample of newborn children, and randomly assign half of them to the fat condition and half to the no-fat condition. Children in the fat condition would be required to consume an excessive amount of food throughout their lives. In the no-fat condition, the diet would be the same, but the amount of intake would be carefully restricted. Each child would have to be monitored to be certain that no one was breaking the rules. Except for the difference in caloric intake, both groups of children would have to be treated in the same way—study, exercise, emotional pressures, etc. Then, if we found the tubby types dropping off early, we would have a good basis for proclaiming "Fat is fatal." Is this study likely ever to be conducted in human subjects? No way. If not for the ethical reasons that prevail in a democracy, the logistical problems would disqualify it even in a dictatorship. Thousands of subjects would have to be assigned to each condition in order to establish stable death rates in later years. Each child would have to be watched almost on a one-to-one basis throughout life. The national budget would reel under the various demands of this one study.

Reduce That Pressure, Clean Up the Environment

"Well," you say. "There's one fact that is indisputable. During the past eighty years or so, the rate of deaths due to cardiovascular disorders, cancer, and what have you, has been on a steady increase. It's got to be due to the pressure cooker we live in these days, food additives, pollution, and what have you."

No doubt about it. The death rate in such dreaded disorders as cancer or cardiovascular disorders has generally been on the rise since the turn of the century.* This can be seen in figure 12.1. In contrast, deaths attributable to communicable diseases, such as tuberculosis, pneumonia, and influenza, have been on a seventy-meter ski jump. The drop in these diseases is easily accounted for: It's due to the development of so-called miracle antibiotic drugs. But are we free to conclude that the increases in cardiovascular disorders and cancer are the inevitable consequences of our age of anxiety? Work and political stress, deteriorating environment, atomic bombs, and nuclear reactors? Once again, the fact that these events may have increased together does not prove that one causes the other. I feel there is a far more reasonable explanation: Cancer and cardiovascular disorders are primarily diseases of the middle and late years of life. Life expectancy has increased so much since 1900 that a greater proportion of the population is in these age categories. A white male born today can expect to live about 45 percent longer than one born at the beginning of 1900. The white female life expectancy is about 50 percent longer (see figure 12.2).

*However, there has been a dramatic decline in strokes during the past few years.

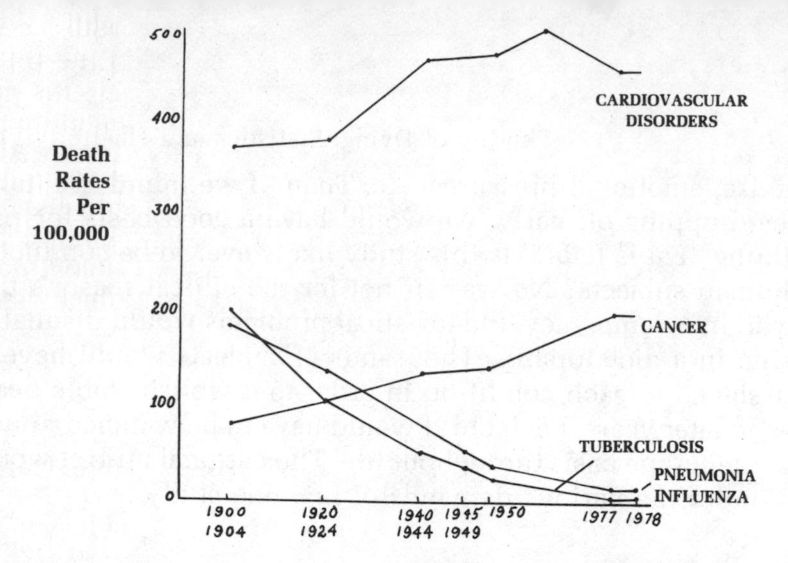

Fig. 12.1 Death Rates. This graph shows death rates, per 100,000, since 1900 among 4 leading categories of diseases. Note that deaths attributable to cardiovascular disorders and cancers have generally increased while deaths due to TB, pneumonia, and influenza have declined dramatically.

By living longer, we're reaching age groups in which we're more vulnerable to cardiovascular diseases and cancer. Figure 12.3 shows five different age groups and the proportion of deaths attributable to each of these illnesses within each age group. A startling finding emerges from this welter of statistical facts: The age groups 55 and over account for 91 percent of all cardiovascular deaths and 81 percent of all cancer deaths. Many of

Fig. 12.2 Life expectancy of male and female white and non-whites born during various years of this century. Note that the life expectancy of a white male has increased by 70.00 − 48.23/48.23 × 100 = 45.14%; for white females, it is 77.70 − 51.08/51.08 = 52.11%. For non-white males, it is 98.52% and a whopping 108.62% for females.*

*Based on *Information Please* Almanac for 1980 (New York: Simon and Schuster, 1979).

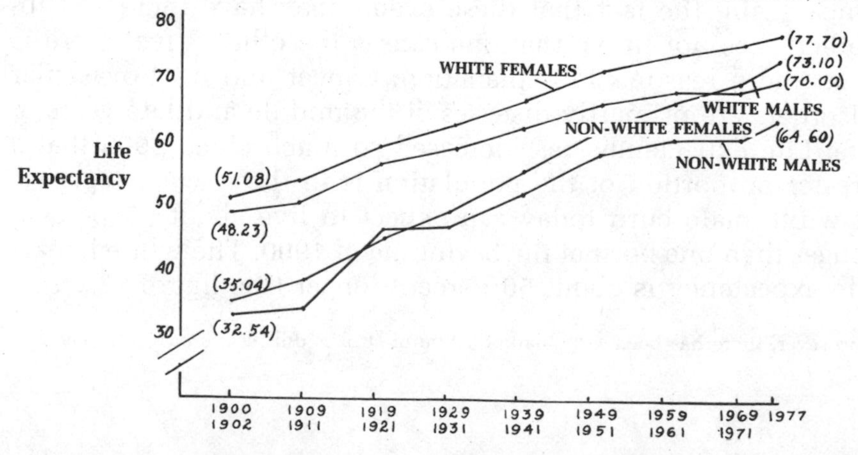

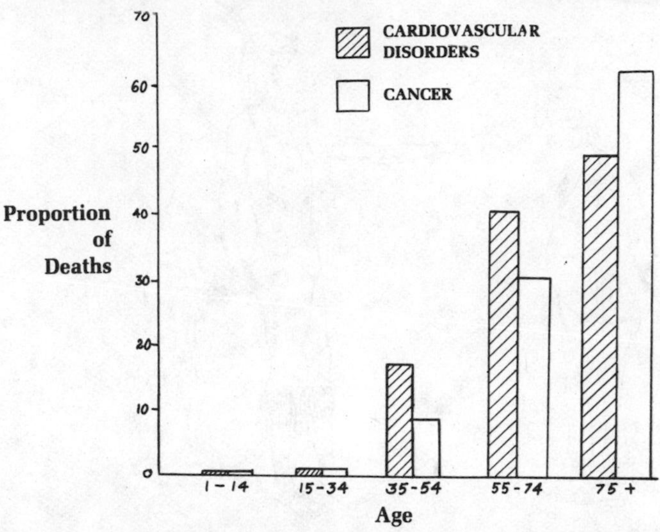

Fig. 12.3 Death Rates by Age. The proportion of deaths attributable to cancer and cardiovascular disorders among five different age groups. Cardiovascular disorders claim fewer than 1% of their victims prior to 35 years of age; cancer claims just under 2%. Note also that about 65% of cancer deaths occur among people 75 years of age and older.*

* Based on data appearing in *Ca—A Cancer Journal for Clinicians*, Vol. 29, No. 1. (January/February 1979).

our grandparents and great-grandparents never ascended to those lofty ages. Had they done so, I have no doubt that most would have succumbed to cancer or cardiovascular failures.

What About the African?

"But what about the African?" you ask. "Everybody has heard about those Africans. They rarely eat animal fats, and heart and vascular conditions are practically nonexistent among them. Doesn't that prove something, Mr. Skeptic?"

"You're talking about saturated animal fats, cholesterol, and hardening of the arteries, I presume?"

"You got it. We all know that most animal meat is high in saturated fats. Americans eat meat, and Americans have high levels of cholesterol. They also have a lot of heart attacks. This is so generally known, why, even my doctor has heard about it."

"What do you know about the life expectancy in Africa?"

"Oh, is it that age thing again?"

"That's exactly what it is. There are many places in Africa where a person's considered old at the age of thirty-eight. They

"Saturated animal fats, cholesterol, hardening of the arteries, vascular conditions—not me, bubbie!"

die from pestilence, famine, drought, and from parasitic diseases I can't even pronounce. They don't live long enough to experience the good fortune of succumbing to hardening of the arteries, heart attack, or coronaries."

To complicate matters even further, the medical profession cannot even agree that high levels of animal fat in the diet inflict any harm. In the United States, the average intake of fat is about five ounces per day. The Japanese average about one tenth this amount. Do they appear to benefit from this extremely low cholesterol diet? Yes and no. While the incidence of coronary disease is low, such disorders as hardening of the arteries in the aorta and in the brain, diabetes, and hypertensive vascular disease are as common as in the United States, and perhaps more so.* Once again, finding that two things are related does not establish a causal link between them. Well, is it precisely that type of evidence (correlational) that has incriminated the cigarette companies?

Are the Cigarette Companies Right?

For years, the cigarette companies have been like voices crying in the wilderness. They have sounded like the original broken record, repeating over and over again, "Nobody has proved that

* Beeson, P. B. and McDermott, W. Textbook of Medicine. Philadelphia: W. B. Saunders Co., 13th Ed., 1971.

cigarette smoking causes lung cancer in human beings." Are they right?

Strictly speaking, they are. The evidence incriminating cigarettes in lung cancer is of the correlational type. Among the relationships already established are the following:

1. The greater the number you smoke per day and the longer the smoking habit has endured, the greater the likelihood of contracting lung cancer.

2. The relationship holds in urban and rural areas, for different racial and ethnic groups, and in countries scattered all over the globe.

The problem with correlational data is that they do not rule out competing explanations. It is possible, for example, that smokers differ from nonsmokers in many ways other than being hooked on the habit. They may differ in personality, dietary habits, biological characteristics, or in their susceptibility to lung cancer. While all of these may seem far-fetched, they exist as possibilities. As long as they exist, a definitive statement, "Cigarette smoking certainly causes lung cancer," will not be made by most scientists.

But let's not underestimate the force of the correlational data. Shown below are two figures prepared by the American Cancer Society. Figure 12.4 shows the death rates per 100,000 males attributed to various types of cancers since 1930. Note the sophistication of the ACS. The death rates are age adjusted in accord with the 1940 census. This means that the increased life expectancy in this country until 1940 has been taken into account. Note that, with one notable exception, deaths caused by all forms of cancer have been either stable or decreasing. The exception? You guessed it. Lung cancer. More than a ten-fold increase in a few short decades.

The women present a similar case except, starting to chain-smoke much more recently than men, they have a lot of catching up to do (see figure 12.5). They can now boast that their death rate due to lung cancer equals that of men in the late 1940s. You've come a long way, baby, but still have a way to go!

True, even this evidence is not conclusive. And the fact that cigarette smoke has increased the rate of cancer in laboratory animals does not prove the case either. It *proves* only that cigarettes are bad for the health of laboratory animals. So what's a person to do?

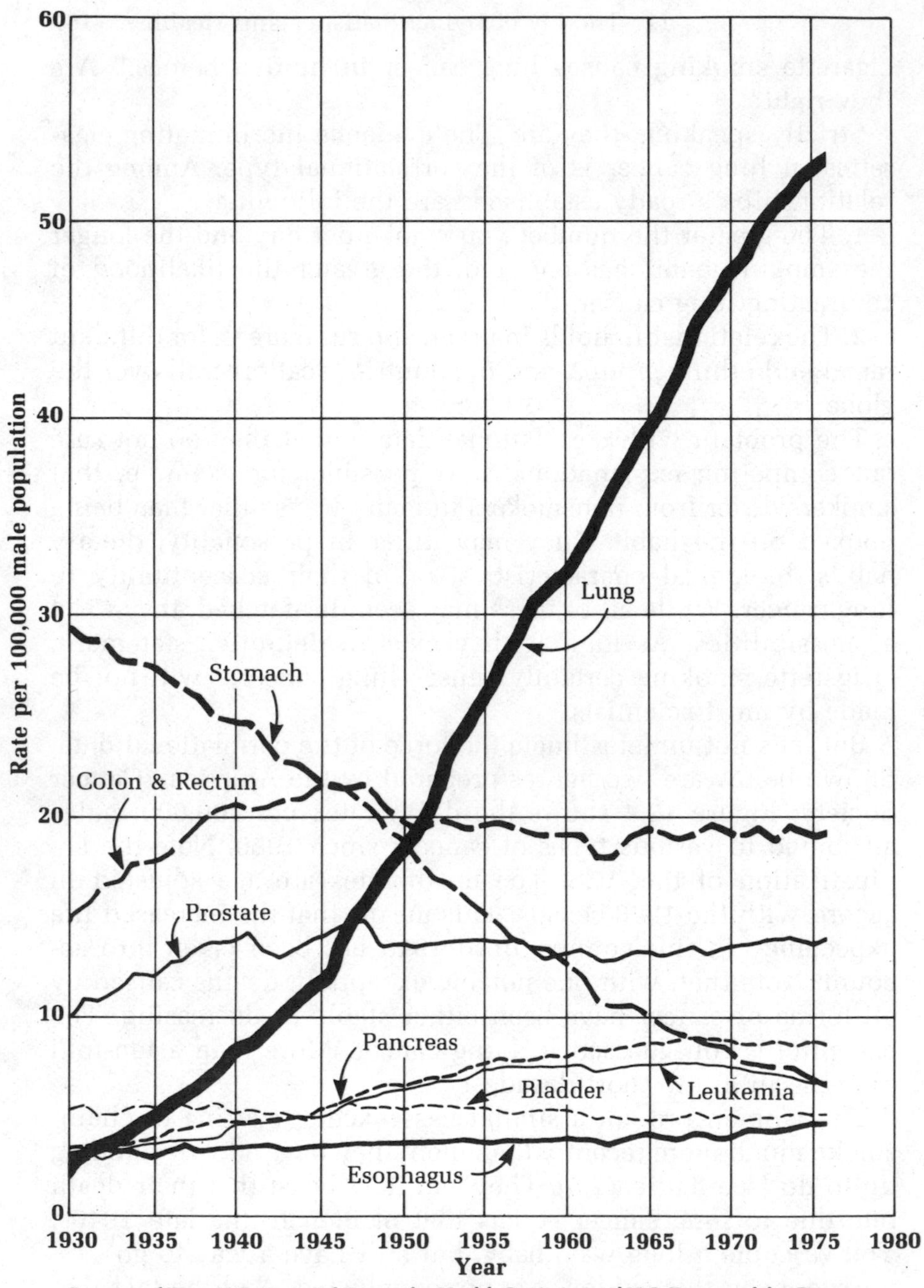

Sources of Data: U.S. National Center for Health Statistics and U.S. Bureau of the Census.
*Standardized on the age distribution of the 1940 U.S. Census Population.

Fig. 12.4 Age-adjusted Cancer Death Rates* for Selected Sites, Males, United States, 1930–1976

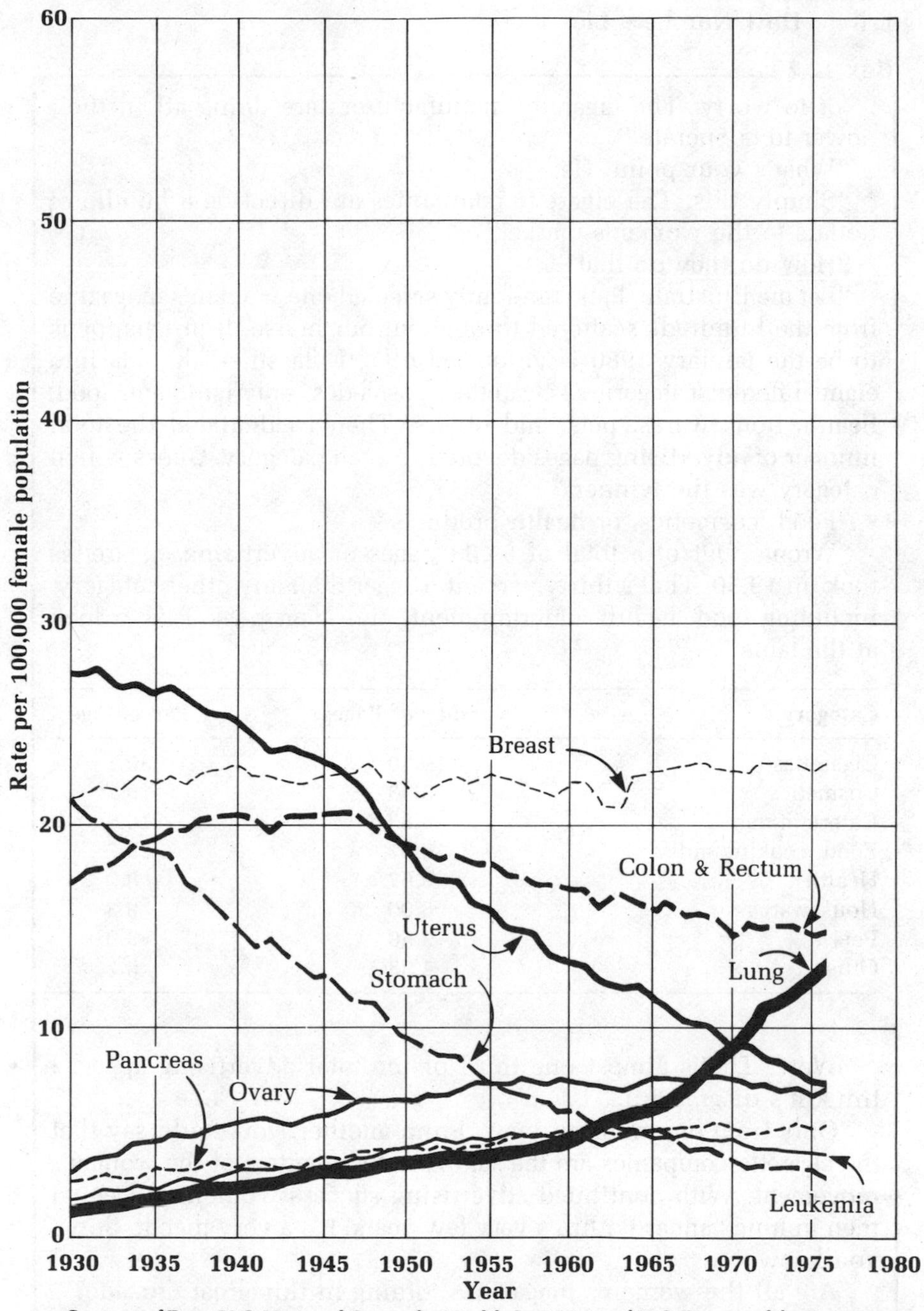

Sources of Data: U.S. National Center for Health Statistics and U.S. Bureau of the Census.
*Standardized on the age distribution of the 1940 U.S. Census Population.

Fig. 12.5 Age-adjusted Cancer Death Rates* for Selected Sites, Females, United States, 1930–1976

Box 12.2

"Not to worry. The cigarette manufacturers are doing all in their power to cooperate."

"What's your point, Hal?"

"Simply this. The cigarette companies are directing a bundle of dollars to the women's market."

"How do they do that?"

"Let me illustrate. I just randomly selected one woman's magazine from the hundreds scattered throughout our house. It just happens to be the January 1980 issue of *McCall's*. I classified the ads into eight different categories—cigarettes, cosmetics, entertainment, food, health, housewares, pets, and others. Then I calculated the total number of advertising pages devoted to each category. Guess which category was the winner!"

"Food, cosmetics, or health products."

"Wrong. Out of a total of 64.83 pages of advertising, cigarettes took up 19.50. That's thirty percent. Bigger than any other category, including food, health, entertainment, and even pets. Take a look at the table."

Category	Number of Pages	Percentage
Cigarettes	19.50	30.1
Cosmetics	3.33	5.1
Entertainment	6.83	10.5
Food, cooking aids	16.67	25.7
Health	5.67	8.7
Housewares	6.00	9.3
Pets	2.50	3.9
Other	4.33	6.7

"Wow! That's almost one third of the total advertising space. I think it's disgraceful."

"Only from one point of view. From another, you could say that the cigarette companies are the staunchest supporters of the women's movement. With continued advertising success women will equal men in lung cancer within a very few years. It's a very macho thing, you know."

"Are all the women's magazines joining in this great crusade?"

"No. *Good Housekeeping* is the party-pooper. It doesn't take cigarette ads."

"Well, I'm certainly not going to let any daughter of mine read *that* magazine."

"In that case you'll love *Psychology Today*. It is doing all it can for the physical and mental health of its readers. Only 40 percent of the full page ads in the May 1980 issue push booze and butts."
"Now that's *my* type of magazine."

For my part, I find the alternative explanations of probable links to lung cancer unconvincing and contrived. Smoking seems the best candidate on the horizon at this point in time. In fact, my wife and I found the correlational evidence so convincing that, on January 2, 1975, we both cold-turkeyed the weed. We haven't regretted it a moment and we have made our dentist very happy.

The World Behind Medical Statistics

Some years ago, B. F. Skinner and a colleague took some hungry pigeons and placed them individually in small cages.* Then, on an irregular and unpredictable schedule, food was delivered to the pigeons. When the experimenters returned hours later, they found that most of the pigeons had developed elaborate and repetitive behavior patterns they labeled as superstitious behavior. Some would turn clockwise, others counterclockwise, and yet others would execute courtly bows. Apparently following one of their everyday and normal movements in the cage, a food reward had been fortuitously delivered. This reinforced or strengthened the behavioral pattern. A number of accidental pairings of the behavior and reward was sufficient to establish a "superstitious" response.

Is this much different from the superstitious behavior many of us acquire? The baseball player didn't shave on the morning of a particular game and he subsequently rapped the game winning hit. He stopped shaving for awhile. The gambler stubbed his toe on the day he hit big. He is now known as Limpy Louey. Of course, being a rational person, I am completely free of su-

* Morse, W., and B. F. Skinner. A second type of "Superstition" in the pigeon. *American Journal of Psychology*, 1957, 70, 308–311.

perstitious behavior. What's that neighbor? You noticed I never wash the car on the day of a family outing? You crazy? You think I want it to rain?

But how does all of this bear on medical statistics? For one thing, the medical literature is clogged with reports that are based on casual observations rather than preplanned, carefully thought out, and meticulously executed experimental research. This is one of the reasons that one doctor will tell you to apply heat to a bruise while another will advise cold. Both treatments may seem to be effective because most of us recover from most illnesses without benefit of doctor or drug. Our own body defenses, evolved over eons of time, charge to the rescue most of the time. But, if we happen to be receiving a treatment at the time we recover spontaneously, we often attribute the recovery to the treatment. We have all heard the bromide, "Treat a cold with the entire arsenal of modern medicine and you will recover in a week. Otherwise it will take seven days."

Even an occasional patient will recover spontaneously from an "incurable" illness. Some call it a miracle but others will bestow credit on whatever is being done to the patient at the time. If he or she happened to be receiving an extract from an apricot pit, laetrile is credited. Pay no heed to its failures. Is it not at least remotely possible that some people recover from cancer because their ancestral line has evolved effective defenses?

Then there is the placebo effect. As we see in chapter 13, some people react favorably or unfavorably to a drug or treatment because they *think* it will have an effect. A pill containing an inert ingredient will appear to cure some people while causing others to break out in a rash. Modern medical and behavioral scientists have little more than a glimmer of an understanding of the "powers of suggestion," also called mind over matter.

Finally, the insurance companies are little help. They sponsor a great deal of research of the correlational type. After all, their primary interest is not in saving lives but in discovering factors related to foreshortening life. In other words, they are not looking primarily for causes and cures. Rather, they are interested in uncovering indexes that predict life span. If they were to find that the length of the little toe is negatively related to life expectancy, we would soon find insurance doctors measuring your little toe. "Sorry, but your little toe is too little. We must put you in a special risk category."

Insurance companies publish like it is going out of style. Moreover, they are constantly sending summaries of their findings to the media. How many reporters, reading that little little toes and short life go together, can resist the temptation to write: "Short little toes can *cause* premature dying"?

CHAPTER **13**

You Can Prove Nothing Safe

Guess Which Hand

Do you like party games? You don't? Oh, damn, just when I was going to show you one that is sure to mystify your friends. You sure you don't want to hear? You do? Great!

To start out, find two brand-new coins of the same year and denomination. Be sure that neither has a distinguishing mark or scratch that would give it away. Keep them always on your person so that they are available to give life to a dull party.

Then say, in a loud voice, "Well, my broker is E. F. Hutton." If we can believe the TV commercials, everybody will freeze in place waiting for the pearl of wisdom to drop from your mouth. Then say, "I have developed a mysterious power that allows me to control the minds and activities of other people." Then a scene like this might follow:

"Come on, now," says Matt, "you're putting us on. Tell us what E. F. Hutton has to say. You know, I've never heard them say a thing."

"I don't give a damn what E. F. Hutton has to say," interrupts Fran, as everything in the room is suspended, tense, and hushed in response to the name of the magic brokerage house. "I want to hear about this mysterious power."

"I told you, he's putting us on. There's no such thing as mind control."

"Yes," you answer. "It's a branch of extrasensory perception known as telepathic kinesis."

"Aw, come on," Matt says.

"Give him a chance, Matthew."

"Here's what I'll do," you say, removing one of the coins from your pocket. "I'll conceal this coin in the palm of one of my hands. I'll ask you to guess which hand. Here's where the mind control comes in. I will control your mind so that you always make a wrong choice."

"Aw, come on," Matt says.

"No, really. Here, Matt, look at the coin."

"New 1981 penny."

"Good. Now I'm going to put my hands behind my back and conceal it in one hand. You try to guess which one."

While all of this has been going on, you have concealed the second coin in your other hand. You place your hands behind your back and deposit one coin in each hand. You bring your fists forward and ask Matt to make a choice.

"Your left hand," Matt says.

You reply, "Oh, I'm sorry. You're wrong," while displaying the coin in your right hand.

Do this again and again. Before long, you'll either send someone scurrying for asylum from your thought control or they'll catch on. You really never know what is going to happen. Children under six are particularly good victims. Some will persevere for hundreds of trials.

If you look carefully enough, you'll find a lesson in this little party game. Namely, it is exceedingly difficult to prove something by a process of elimination (in other words, if a coin is in the right hand, it cannot be in the left). Many people expect

Which hand is it in?

science to prove that something—a dye, a pesticide, a food preservative, a drug such as pot or LSD—is not in the left hand (harmful) by showing that it is in the right hand (safe). In this chapter, I will show that this is a forlorn and hopeless expectation from either scientific or statistical analysis. The grim and inescapable fact is that it is relatively easy to demonstrate that something is harmful, but impossible to prove that it is safe. Marijuana will not someday be legalized because it has been proved safe. It will become accepted when and if the weight of evidence suggests that its harmful effects appear no greater than many other poisons we ingest or inhale daily.

But Dodoes Are Safe and Effective

Surely you've heard commercials on television that boldly proclaim: "Dodoes have been proven clinically safe and effective in the treatment of nagging back ailments," and quite possibly you accepted this statement without much question. The truth of the matter is, however, that nothing can be proved to be safe. You can prove things unsafe, but the opposite—proving them safe—is not in the realm of possibility.

Now that's a pretty sweeping statement and I guess I had better be prepared to defend it. So let's get to it.

Let's start out by using a drug as an example, and see how we might evaluate its effectiveness. Let us imagine that the drug is supposed to be effective in the treatment of pains coming from the lower back. You identify a sample of people with complaints of chronically aching backs. You randomly assign half the patients to one experimental condition and half to the other. You also use a double-blind study (see box 13.1) in which neither you nor the patient knows what treatment is being administered. Of course, half the patients receive the drug and half the placebo (an inert ingredient that appears to be the same as the drug). Subsequent testing reveals that the individuals getting the drug actually experienced relief from pain in the lower back, whereas relatively few of the patients in the placebo, or control, group evidenced similar relief. Moreover, an analysis of the probabilities forced you to the conclusion that the greater improvement rate of the experimental subjects was not likely

Box 13.1 The Placebo and Double-Blind

"Hold it a minute, Dick."

"Yes?"

"Don't you think an explanation is in order? Otherwise some readers will think that researchers using the double-blind technique are engaging in some cruel and barbaric practice. Don't forget how the Army doused San Francisco with bacteria and dropped bacteria-loaded light bulbs in the New York subway system. I want to assure them that the double-blind technique is a perfectly reasonable procedure for collecting medical data."

"Go ahead, you've got the stage."

"Thanks. Let's start with the placebo. It's an indisputable fact that the health of some people will change because they *think* they are getting something that has medicinal value. Swallowing an empty capsule will cause some ill patients to get better. Some will even get worse if they think that they are allergic to pills. Both getting better and getting worse without the benefit of an active medical ingredient describe the placebo effect."

"But why use a placebo?"

"Simply to distinguish the effect of the drug from the effect of the placebo. If fifty percent got better following administration of the drug and a similar percentage respond favorably to the placebo, we can't get too excited about the curative effects of the drug."

"Ah, I see. And I suppose the patient does not know what is being administered?"

"Precisely. That's half of the double-blind. The patient is blind to what he or she is receiving."

"And the other half?"

"The researchers don't know what is being administered. They are also blind."

"Now wait a minute. That strains credulity. Sounds like a governmental agency where nobody knows who's doing what to whom."

"It's not that bad, Dick. The active ingredient . . . the drug . . . and the placebo receive coded identification at the beginning of the research. The researchers simply administer a coded capsule, injection, or what have you. We don't want their own biases concerning the effectiveness of the drug to cloud their later observations and diagnoses. When the study is completed and all the data are in, the code is broken. At this point we can distinguish placebo from drug effects without fear of contamination by placebo effects or researcher bias."

"Wasn't that done with the Salk polio vaccine?"

"Yes, and thousands of studies since then."

to be due to chance factors. To your satisfaction and that of fellow scientists, you have proved the drug effective. But that's the easy part. How do you go about proving it safe?

This problem, friends, is an entirely different ball of wax. Let's start out by asking the big questions: Safe for what? For whom? Within what time frame? At what dosage level? For what body system? How would you go about determining that something does not have any undesirable effects on any body system, either immediately after administration or at some unspecified future date? I'm waiting for your answer. Good. Basically what's done is that you look at the prior history of drugs of this sort and develop informed hunches about the locus of possible harmful effects. You look for unfavorable effects in certain physical structures, or certain enzyme systems, or certain aspects of behavior. You must be selective, for reasons that will soon become obvious. You then set up a series of studies in which you follow your hunches. You look at the system or systems that you feel are most vulnerable to unfavorable reactions, if indeed there are any. You set up a double-blind experiment in which the drug is administered to the subjects at varying dosages, and then you monitor the bodily or behavioral system that you have selected for study. If you find that the difference in the number of undesirable side effects among the experimental and control subjects is no greater than would reasonably be expected by chance, you feel confident that the drug produces no untoward effects.

So far, so good. But have you proved that the drug is safe? Not by a long shot. Typically, failure to find evidence of deleterious effects is taken as evidence that the drug is safe. However, you have selected only a relatively few out of uncounted, and perhaps uncountable, systems for study. As we have noted, you made an informed guess. Your guess was that if the drug has bad effects, these effects will manifest themselves in system A or system B or behavior C or something of this sort. Your conclusions must be restricted to only those systems that you have tested, the dosage levels that you used, and the time frame within which the effects were studied. So you haven't really proved that the drug is safe in any general sense of the word. You have merely established that individuals getting the drug did not appear to have any untoward side effects when compared with individuals getting the placebo.

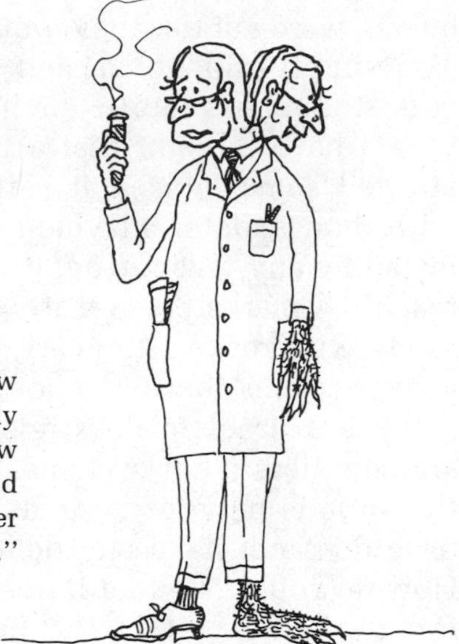

"Doctor, our experiments show there are unusual and potentially harmful side-effects to this new drug. We believe its label should reflect certain consumer cautions."

But then another question begins to intrude into your thoughts. It is possible that the drug has an adverse effect on only a small segment of the population and that this effect could not be detected in a sample of the size that you used in your study?

When you say that a drug is safe with respect to certain physical systems that you studied, what you really have said is that you haven't detected any harmful effects among the sample of subjects that you've chosen. But what if only one in every thousand people suffers a serious side effect? This fact would be completely obscured in the original research study. There would be no way of knowing it. This rare type of harmful effect will come to light only after many thousands of people have actually taken the drug in private use. Remember the thalidomide disaster. Witness also the birth control pill. Ten years ago, safe; five years ago, safe for most women; today, caution and frequent physical examinations advised.

But let's go a step further. We've examined only a few out of possibly a million different systems. For example, there are several thousand enzymes operating in the body. These enzymes control all aspects of our biochemistry. They are the precious templates that help chemicals to line up in such a way that reactions can proceed efficiently. What if the drug produced an effect in an enzyme system that you weren't studying? If the effects were subtle, they would probably go undetected. Why? You simply did not design measurement of that system into your study. Let's return a moment to the drug thalidomide. Who would have thought that a tranquilizer would have adversely affected some fetuses at certain stages of development? Now with thousands of enzyme systems that could be adversely affected by any given drug, it is obvious that no drug could ever reach the marketplace if it were required that we check out its safety with respect to every known enzyme system. And what about systems that have not yet been discovered?

But our problem does not end with enzyme systems. There are countless other systems operating in the body. What about the various nervous systems? How do you demonstrate that a drug does not have any deleterious effects on neural activity? How do you prove that it doesn't have an effect that is selective for a specific part of the brain? The brain contains literally billions of brain cells and clusters of cells whose functions are largely still unknown. And this is only the beginning.

What about various aspects of behavior? A drug may not affect motor coordination, but maybe it has some effect on memory, or maybe it doesn't have an effect on memory in general, but on some very selective aspect of memory. Or maybe it has an effect on the ability to solve abstract mathematical problems. How could one possibly study all the behavioral dimensions that might be adversely affected?

I've only begun to mention various systems of the body that could be adversely influenced by the drugs. So anybody who says that something has been proved safe is telling a lie. It may be told in ignorance, but it's still not true. There is no way that anything—not even water—can be proved safe. All we can say is that given the sample of subjects and the measures that we chose to study, we did not find any evidence of undesirable effects. That's as far as we're entitled to go. This is not saying that there is anything wrong with science. And this is not saying that there is anything wrong with medicine or with the pharmaceutical houses. I am simply stating that it would be an impossibility to study all conceivable reactions to any given drug, or food additive, or pesticide, or chemical agent of any kind.

If you don't object to beating a dead horse, let's go one step further. Let's make up a figure. Let's say that there are 100,000 different systems in the body that are sufficiently distinct to be independently affected by a chemical agent. Now one might say, "All right, if you're really very, very careful and you have several lifetimes to spend on research, you might conceivably study a hundred thousand different bodily systems to find out whether or not a given drug is safe or, more precisely, that it has not produced evidence of any undesirable effects."

Now you run into the next problem. If you study a given system in isolation, you may not find any evidence that that system is unfavorably affected by a given drug. However, there is a concept that's very important in the field of statistics. It's called *interaction*. A given variable operating alone may produce one effect, but when functioning in the presence of a second variable, the effect may be quite different. In other words, although A equals one and B equals two, A plus B may not equal three. Take a simple illustration. You put fertilizer on your outside plants during a period of severe drought. The effect of the fertilizer is to burn all of the microfine root structures. Your plants die. But the same fertilizer administered when moisture is available causes the plants to absorb the nutrients in the fertilizer—instead of dying, the plants thrive. In other words, the effects of fertilizer administered without water are quite different from those of fertilizer administered with water. And you can get very complex types of interactions. For example, the effect of variable A may be dependent on the presence of variable B, but the effect of both may be dependent on C,D,E,F, or G. And if you have 100,000 systems operating in the body, it is quite possible that these systems are interacting in billions of different ways.

Again I state quite categorically, there is no way of studying all these various interactions. Now let me illustrate with another example. As we are all aware, there are many additives placed in food. The claim is made that a given additive is safe. How is it found to be safe? As stated before, some astute researcher tested a series of different hunches about systems that could be sensitive to the undesirable effects, if there were any. But typically each system is tested in isolation of every other system. But what if A alone is safe with respect to a given system, and

B alone is safe with respect to a given system, but A and B together are lethal? Well, we're all familiar with the effect of barbiturates and alcohol. You take a barbiturate as a sleeping pill and it puts you to sleep. It has its desired effect. You take alcohol alone and alcohol is also a depressant; it makes you sleepy and you go to sleep. You take sufficient alcohol and barbiturates together and the sleep is sort of permanent.

One more example. There are many additives placed in the things we drink. For example, you wouldn't believe the line-up of additives that is put in beer. Some cause the beer to maintain a head for a longer period of time, others make the color lighter or darker, others halt bacterial growth, others maintain the flavor, and still others keep the bubbly going until there's no more beer to bubble. There are different additives for just about every effect that you could ever want. Why someday, they'll remove the beer and you'll drink the additives and notice no difference!

One additive that was rather thoroughly tested is a chemical called cobalt sulfate. It is used in some European beers to extend the life of the foam over a longer period of time. The only trouble is that the tests for toxicity were conducted in isolation of the very beverage—beer—to which it is added. It so happens that alcohol plus this cobalt sulfate interact in such a way that they form a deadly poison for some people. Not for many, but for some. Over the years this combination has been incriminated in the death of at least fifty beer drinkers. That is why the labeling of the ingredients of all domestic and imported alcoholic beverages became mandatory on January 1, 1978.

Alas, I can no longer quaff a little booze without putting on my bifocals.